Margarita Ryutova-Kemoklidze

The Quantum Generation

The Quantum Generation

Margarita Ryutova-Kemoklidze

Highlights and Tragedies
of the Golden Age
of Physics

With 17 Illustrations

 Springer

Margarita Ryutova-Kemoklidze
Budker Institute of Nuclear Physics
630090 Novosibirsk, Russia

Translator:

John Hine
22 Beaumont Road, Bournville
Birmingham B30 2DY, Great Britain

Title of the Russian original edition: M. P. Kemoklidze, *Kvantovnyi vozrast.*
Seriya: Istoriya nauki i tekhniki. Nauka, Moscow, 1989.

Artwork:
Figures 13 and 15 are drawn by Roberto Bartini and the remaining 15 line drawings by Efim Bender.

ISBN 978-3-642-49359-1   ISBN 978-3-642-49357-7 (eBook)
DOI 10.1007/978-3-642-49357-7

Libary of Congress Cataloging-in-Publication Data.
Ryutova-Kemoklidze, Margarita, 1939– [Kvantovyï vozrast. English]. The quantum generation:
highlights and tragedies of the golden age of physics / Margarita Ryutova-Kemoklidze.   p.   cm.
Includes bibliographical references (p.   ).
ISBN978-3-642-49359-1
1. Physics – History – Popular works. 2. Quantum theory – Popular works.   I. Title.
QC7.R9613 1995   530.1'2'0922–dc20   94-36678   CIP

© Springer-Verlag Berlin Heidelberg 1995
Originally published by Springer-Verlag Berlin Heidelberg New York in 1995
Softcover reprint of the hardcover 1st edition 1995

Typesetting: Data conversion by Springer-Verlag
SPIN 10018796   56/3140-5 4 3 2 1 0 – Printed on acid-free paper

Foreword

Quantum mechanics dates its anniversaries from 1925, the year when its first versions, matrix and wave mechanics, were born. The quantum itself had by that time reached the age of 25, having first seen the light of day on the eve of the nativity of the 20th century itself. Either shunned or completely unnoticed, the quantum seemed for a long time to be an illegitimate child. Even Max Planck himself, its own father, could not bring himself to accept his monstrous child until he was forced to do so – that is, until it became clear that the problems which were arising in physics could not be solved without the quantum.

At first the quantum "like a greasy stain, soaked through all the different branches of physics" and then, like an explosion, came the creation of quantum mechanics. The overwhelming majority of those who first created quantum mechanics were just about of an age with the quantum itself. These were very young people, born as the new century was born, give or take a year or so. Hence the phrase was coined – "the quantum generation". At that time there was quantum everything – towns, steps, park benches, lodging houses and so, naturally, the quantum generation.

Many books and articles have been written about the creation of quantum mechanics, about its youthful creators, about the way it re-shaped the long-accepted laws of physics. In the early sixties the California State University at Berkeley set up an Archives Committee with the job of collecting all the surviving information – scientific, psychological, social and domestic. If all that has been written about quantum mechanics is collected, the result will be an enormous library.

It is not claimed that this book, which is also concerned with the life and times of the quantum generation, fills in any gaps in the record. Its justification, perhaps, lies in the fact that it contains a measure of "surviving information" in the form of the eye-witness accounts of one who lived through those momentous events – Professor Yuri Borisovich Rumer. Yuri Borisovich was one of the contemporaries of the quantum. He lived in Göttingen, in Frau Grönau's "quantum lodging-house" and was cooked in the then brand-new "quantum kitchen".

Albert Einstein's correspondence with Max Born was published in Russian in the 1970s. Not long before his death Max Born decided that this correspondence, which continued from 1916 to 1955, right up to Einstein's death, and which had never been intended for any other eyes or any wider audience, should be prepared for publication. He added explanatory notes practically to every letter. Several of the letters could not fail to be of special interest to the Russian reader. In these we hear of a certain young man from Russia:

> "Dear Einstein,
>
> A young Russian turned up here recently. He has a theory of six-dimensional relativity. The young man's name is Rumer. If his work impresses you favourably I would like to ask you to help him – – " [1, p 101]

Born was constantly asking Einstein for help. In those days there were no salaries for scientific work and able young people often found themselves without means of support. The foundations could not provide grants for all applicants and Einstein's support was very important. All those around whom the young people flocked wanted his support – hence Einstein's caution: "If, just once, I slip up and recommend a second-rate physicist, I shall lose all my influence and shall no longer be able to help anybody. I am painfully aware that I behave like a horse-trader, praising my wares, pointing out the fine teeth and the brisk pace." [1, p.129]

Rumer turned out to have the right kind of pace. On the 14th of December, 1929, Einstein wrote to Born:

".. I took to Mister Rumer ..." [2, p.15]

Yuri Borisovich loved to talk about his life. Not so much about himself as about the people he had met. His stories were lively and short, the genre lying somewhere between the anecdote and the parable. He would often switch from character to character and from scene to scene in the middle of a discourse. But that did not matter. The main thing was that he conveyed to the listener the feeling of the age and the spirit of the generation which grew up with the century. One saw living pictures and living characters. One saw the Moscow lad from Maroseika who had been there when the old gas lamps were replaced by electric lights, who had sat in freezing lecture-rooms, who had religiously divided the cold pearl-barley porridge into equal shares during the Civil War. Often one had to ask the same question several times. For example:

"Yuri Borisovich, please tell me about Born's seminars."

"Born's seminars? Nothing unusual. We used to have cakes brought in. By the way, did you know that certain quantum problems can be formulated in terms of the operators of non-compact $SO(2,1)$ algebra rather than in the language of Heisenberg's algebra?"

There would follow a detailed account of how, this being the case, it was possible to set up an elegant classification of the potentials in Schrödinger's equation. The next time I asked the same question I might learn the best way to teach statistical physics to undergraduates. Or I might ask:

"Yuri Borisovich, tell me a bit more about the way you got to know Landau".

"You know how it was. It was in Berlin".

After several attempts the scene would be sketched out.

"Come to the meeting at the university tomorrow. I will introduce you to a certain young lad from Russia," Ehrenfest had said.

It was December, 1929. The University of Berlin. One of the regular meetings of the Physical Society. It was a big auditorium resembling an amphitheatre. There were undergraduates, postgraduates, guests from various places. In the front row were the Nobel Prize-winners: Max Planck, Max von Laue, Einstein, Rutherford, Nernst, James Franck.

Ehrenfest introduced them. "You will like each other – you will like each other very much," he said in his mangled "Ehrenfest-Russisch" as he presented Lev Landau and Yuri Rumer to each other. "This is Landau. He does not bite." But, in fact, the twenty-year-old Landau did bite. Very much so. He was tall, handsome and very slim with thick wavy hair. He had a mocking glint in his dark eyes – and a cutting tongue. Rumer was also slim; very tall, with eager black eyes and wild curly hair. He, however, was courteous and kindly.

"That first time I met Landau I liked him very much", Yuri Borisovich used to say. "We talked about physics, and I was struck by the ease and flexibility of his understanding. His erudition was natural and playful, like the singing of a bird. He was born to be a physicist. We talked about this and that, and I sensed that he had a better grounding than I had, but not much better, I did not suspect that there was more to this meeting. I did not know that fate had brought me into contact with one of the most brilliant minds of the century. That was to be the verdict of history. We met as equals. He was cocky but not conceited. Moreover, I had been to see Einstein the day before and he hadn't."

These words describe the meeting of two kindred spirits, two men with similar aspirations, whose lives were to bear all the marks of this hard century. There were to be amazing meetings, joys and sorrows, flights and falls. Fate smiled on each of them in a different way and in unequal measure. Landau was marked with the happy star of genius and fate did not delay him long on his way to a fitting eminence.

However, She had prepared for him a terrible blow and a lingering, agonizing end.

Yuri Borisovich Rumer would never have allowed anyone to tell his story in parallel with that of Landau. He regarded himself as a common soldier of science, polishing up his brasses.

"All my life, Fortune has spoiled me with wonderful friends", Yuri Borisovich used to say, "All my life I have had friends with greater strengths and gifts than my own. I have had geniuses around me all my life. Maybe I could have been a poet but, having compared myself with the great poets around me whom I knew so well, I could not seriously consider poetry myself. It was just the same in science. As an undergraduate, I had the luck to associate with the best mathematicians. Luzin, for example. Just imagine – Luzin! When I moved over to physics I studied under Born and Einstein, and my best friend was Landau! Even in prison I was with Tupolev and Korolyov."

And wherever Yuri Borisovich went, whatever he did, he always remained true to his profession and to his principles. To the end of his days he retained his gentle humour and extraordinary kindness.

Whenever I tracked down any of his surviving prison comrades he was always delighted to hear that I had managed to meet them. Only on one occasion when I said, doubtfully, that a certain B... might still be alive, Yuri Borisovich replied:

"I don't think it is worth speaking to him. Out of the 126 inmates in our last prison, he was the only one who was embittered – and, if he's alive, I expect he's still bitter".

Our talks often ended with a poetry-reading – Yuri Borisovich knew an amazing number of poems by heart – or with a heated argument about contemporary theatre. Every time I spoke to Yurii Borisovich I came away with something new. It was all priceless experience, for which I shall always be grateful to him.

In the course of my work on this book I was helped by two of Yuri Borisovich's friends – Lazar Aronovich Lusternik and Rita Yakovlevna Rait.

Lazar Aronovich wrote down several pages of reminiscences as well as a dozen original poems. Unfortunately, the poems, brilliant though they are, would have introduced a digression from the main theme of the book and so are not reproduced here. As for the reminiscences, these were very similar to those already published in his paper "Achievements of the Mathematical Sciences". I have preferred, therefore, (and, indeed, Lazar Aronovich himself advised it) simply to supply references to the published material. I wish also to thank Yuri Borisovich's wife, Olga Kuzminichna, whom I shall remember as a most vivacious and amiable lady, never without a smile and a kindly joke.

I was also helped by A. I. Shalnikov, I. N. Saratovkin and M. M. Zaripov – who all shared their recollections with me. O. Bohr and Y. Kistemaker kindly sent me a book about the Niels Bohr Institute along with some unpublished material about Max Born and James Franck.

I wish to express my most sincere gratitude to all these people. Special thanks are due to Spartak Timofeyevich Belyayev who read the manuscript at various stages of its evolution, gave me an old Göttingen album published before the First World War and took upon himself the responsibility of editing the work. His comments and advice were extremely valuable.

Special thanks go also to Yefim Davidovich Bender who did the artwork – and, indeed, supplied the illustrations before I completed the writing. I am grateful also to E. M. Lifshitz, L. M. Barkov, I. B. Khriplovich, G. I. Surdutovich and L. M. Kurdadze, who read the manuscript and made useful comments. To L. P. Pitaevskii I am indebted for the copy of Y. Mera's book "The Solvay Congresses", presented to him by the author and which he gave to me, in defiance of the childhood rule that gifts are not to be given away". I am likewise grateful to V. G. Zelevinskii for the magnificent book on Göttingen.

I wish to express gratitude to the directors, the librarians and the darkroom staff of the Nuclear Physics Institute for their unfailing assistance.

Finally, I thank my friends, the academic staff of the Nuclear Physics Institute, who sometimes had to lay aside serious scientific matters in order to discuss pre-revolutionary grammar school education or the German uranium project, depending on what it was that concerned me at any particular stage in the writing of this book.

Preface

When I wrote this book I had no idea when, if ever, it would be published.

The central figure around whom the book is constructed is Yurii Borisovich Rumer, who was born at the dawn of this century, the son of a Moscow merchant of the First Guild. He was not destined to live an uneventful life. As a youth he experienced war and revolution. Later, working as Max Born's assistant in Göttingen, he witnessed and played a part in some of the most dramatic developments of modern physics. When the Nazis came to power Rumer returned to Russia. Just at the point in his career when he was ready to produce his finest work he was arrested. He was placed in "Outfit Number One", the "Golden Cage" set up and supervised by Beria himself. Within the confines of this institution the conscientious workers of the KGB brought together the USSR's best aircraft designers – Tupolev, Korolyov and Myasishchev. To assist in the development of jailhouse aeronautics many other eminent academics – physicists, mathematicians, engineers etc. – were also arrested and brought in. Rumer served his time right to the bitter end of his ten-year prison sentence and then spent five more years in exile in the Taiga.

Obviously, when I finished work at the end of 1981 the time had not yet arrived when such a book could be published. Quite apart from the account of Rumer's own dramatic life, there were several other revelations in the book which made it unpublishable. The mere mention of the "Golden Cage", that prison on Radio Street in the centre of

Moscow which outshone all others with the concentrated talent of its inmates, was enough to condemn it. To make matters worse, the book described the persecution of Luzin, the greatest of Soviet mathematicians, and in its pages appeared the letters which Pyotr Kapitsa sent to Stalin and Molotov, knowing as he signed each one that he was signing his own death warrant. A book with such contents hardly seemed likely to win the heart of a Soviet publisher at that time.

However, in the little Siberian town of Akademgorodok, where Rumer then lived and where I wrote the book, everyone knows and influences everyone else. Over the kitchen table, as one might say, I was persuaded by "friends of friends" of the ex-naval officer who worked as chief editor for the Siberian section of the "Nauka" publishing house to submit the book for publication. In similar fashion the ex-naval officer was persuaded to accept it. In February 1982 the manuscript was accepted for consideration by Nauka's" "History of Science and Technology" department. A long saga had begun.

First of all, with some justice of course, Nauka took the view that, as the book was concerned with the life and fate of a certain individual, it should really be regarded as a literary work rather than as a contribution to the history of science. Nevertheless, they did not reject the book, though the ex-naval officer earnestly advised me to accept the fact that no-one would publish it as it stood and that it was better to publish it "in some form" than not to publish it at all. This "form" was to be achieved by cutting out all the facts which might sully the good name of the Soviet state. This was not actually said openly but publication was delayed for a long time while it was pointed out to me that a lot of the information about Rumer's personal life could be deleted and replaced with popular scientific material.

The rewriting of the book dragged on over three years, slowed down by the nature of the editing process. This was no surgical operation which could be finished with one neat excision. Pieces of the "body" were cut out one by one over a long period. Two "surgeons" handled my case. One was a lady suffering from a serious goitre condition who simply would not look me straight in the eye. Catching her eye, however, was easy compared with trying to catch the drift of her real thoughts. Whatever her private opinions, she had an amazing ability not to hear anything that was not in accord with the traditions of Soviet publishing. The other was a buxom lady with a bouffant hairstyle who never doubted for a moment that she was right about everything. These two took it in turns to persuade me to remove such undesirable details as, for example, the story of Rumer's brother Isidor, who was arrested and shot for working with Trotsky.

"Now why do you want to drag Trotsky into it?" said the lady with the goitre, "Rumer left a family – children, grandchildren – you never know how they might be affected. Look, you say that Isidor Rumer translated the theory of relativity from German to Russian. Well then, why not give the reader a brief outline of relativity theory?"

The buxom lady said to me: "Oh, Doctor Ryutova, just ask yourself, why would our reader want to know how Rumer met his true love, how he was arrested at the woman's door and all that stuff? Why not cut it out and write an account of quantum mechanics – wouldn't that be more appropriate?" As she said these things her face shone with an unfeigned enthusiasm, as if the only books she ever read were studies of quantum mechanics by Landau and Lifshitz.

Three years dragged by in this way before the book finally assumed its present shape. I then refused to make any more alterations and "Nauka" refused to publish. The deadlock was broken only when Gorbachev came to power and everybody started to talk about "freedom of speech". This time it was the "Nauki" publishing house in Moscow which took the book on. However, entrenched inertia soon brought the project to a halt once more.

It was all rather like the joke about the train – the one in which the whole USSR is the train and its citizens are the passengers. On several occasions the same thing happens – the train of state is rattling along and suddenly somebody sees that the track comes to an abrupt end just a little way ahead. The different drivers react in different

ways. Driver Lenin, seeing that there are no rails, calls out, rolling his
" r's" : "Stop and rreverse! Yes, definitely rreverse – but not too farr!"
Stalin, the next driver, sees there are no rails and says: "There had
better be rails there tomorrow." Sure enough, the rails are there by the
next day – but both those who lay them and those who watch them
being laid are arrested. Then comes Driver Khrushchev. He looks out
of the window when the train stops and says: "Pull up the rails from
behind the train and lay them in front!" Then Brezhnev. He sees there
are no rails and gives the order: "Draw all the curtains and shake the
train. That will make the people think they are getting somewhere."
Finally Gorbachev. Seeing that the train is at a standstill, he says:
"Open the curtains and let everybody shout all together: 'There aren't
any rails! There aren't any rails!'"

It was at this very moment, when everyone was crying out
"There are no rails!" without making any attempt whatever to do anything about it, that my book set off on its second odyssey. I do not
know how long its wanderings would have continued had it not been
for the intervention of certain academicians; Spartak Belyaev, M.A.
Markov and Lev Pitayevsky got the project moving again and the book
finally saw the light of day in 1989.

I decided not to go back to the original manuscript, which was
almost five chapters shorter than the present expanded version. Three
of the additional chapters are the result of my negotiations with the
ladies from the Siberian branch of "Nauka". These are the first chapter – "By the Sure Path of Experimentation", the second chapter –
"Fiat Lux" and the last chapter – "In the Beginning Was Mechanics".

The two other extra chapters have a slightly different origin, having been written not so much to please "Nauka" as to enable me to
branch out from the life of Rumer to pursue two relevant historical
topics which I personally find fascinating. The sixth chapter is a history of non-Euclidean geometry. This is a part of the history of science
which has always moved me profoundly, a drama which revolves
around some of the greatest minds of all: Gauss, Lobachevsky and

Janos Boljai. The other topic is to be found in the eleventh chapter – "The Choice" which is a study of Werner Heisenberg and the reasons for the failure of the German uranium project. I had been intending to explore this theme separately but then decided to include it in this book as part of the story of Rumer's generation: one of Rumer's friends, Fritz Houtermans, was directly involved in the Project.

In the vast literature devoted to the story of the atomic bomb surprisingly little is to be found concerning Fritz Houtermans and his amazing fate. However, the life of Houtermans deserves to be the subject of a separate book, which I hope one day to write. Meanwhile I have collected some very important material about Houtermans, including his own personal archive which was kindly presented to me by his widow, Charlotte, and his daughter, Giovanna Fjelstad.

I now wish to take the opportunity to add some acknowledgements to those which I made in the original Russian publication. I will start by expressing special gratitude to Ernst Hefter.

The time when the manuscript was on its second lap round the publishing track was not all gloom; there were several bright moments thanks to Dr. Hefter, then physics editor at Springer-Verlag in Heidelberg. Even before the book appeared in Russian Dr. Hefter wrote to me, offering to publish the book. I am immensely grateful to him for this offer and for all the patient and painstaking work he did in preparation for the English version. I am also grateful to Ernst for his warm hospitality in Heidelberg, where I stayed at his invitation in 1990.

On that same trip in 1990 I went to Göttingen for the first time. This town is the setting for so many of the events described in the book and I had been longing to see it for myself. This was a working visit – I gave a talk at a seminar in the Göttingen observatory. In the ancient observatory building, which once belonged to Gauss himself, there is a guest room. That is where I stayed. At the end of the working day, when all the staff of the observatory went home, the old building and I were left alone together. Several of Gauss's possessions are still preserved there – as are the atmosphere and spirit of his time. For me this visit was truly a festive occasion. It was arranged for me by my good friend Franz Kneer, the outstanding astrophysicist and director of the Göttingen observatory.

As the years have gone by my friends in various countries have shared with me and my book all the tribulations we have been through along the road to publication. I am deeply indebted to them for their support.

Duke Guyenne, publisher of the Proceedings on Plasma Astrophysics at the "Piero Caldirola" International Schools of Plasma Physics and unfailing contributor to the Schools was always ready to assist me. He helped me to choose the best title for the English version. Literally, the Russian title is "Quantum Age" .

Translated thus, the title might sound rather grandiose to the English-speaking reader, giving the impression that the book is a definitive study of a whole epoch.

At this point it is my pleasure to thank the book's translator, John Hine. My feeling on first reading his work was that the translation read better than the original. An author who enjoys her own books may seem conceited, but I read the English text with great pleasure.

Wherever I have been, whether to talk to co-workers about our joint projects or to take part in conferences, I have always discussed with my friends and colleagues those issues in the history of physics which concern me most. This has helped me enormously. For this kind of assistance I have to thank my friends Roberto Pozzoli of Milan University, Eric Priest of the University of St. Andrew's, Yoshi Ichikawa of Nagoya and Jun-ichi Sakai of Toyama. I am also grateful to Professor Ter-Haar, whose comments and support were extremely helpful.

My most frequent port of call has been the Harvard Smithsonian Center for Astrophysics in Cambridge and I have always received a warm welcome there. For this I have to thank Gene Avrett, Shadia Habbal, Wolfgang Kalkofen and Stephanie Deeley.

My first American reader and critic was Harold Zirin, director of the Big Bear Observatory in California, whose command of Russian is superb. I am grateful to him for the warm encouragement and kind comments which he sent to me as soon as he had read the book.

I have had some fascinating conversations about the history of physics with Franca and Claudio Chiuderi as well as with Ravi Sudan, Herbert Berk and Dirk Callebaut, which I greatly appreciated.

I wish to express again here my gratitude to my very old friend Arkady Weinstein, who, from the moment the idea for the book was conceived, was always ready to help me whenever I had a problem. In Novosibirsk he read the manuscript at every stage of its evolution and it is to him that I owe my meeting in Minnesota with the widow and daughter of Fritz Houtermans.

Novosibirsk, April 1994   *Margarita Ryutova-Kemoklidze*

Contents

Chapter 1
By *the Sure Path* of *Experimentation*

The mood in physics at the end of the 19th century was one of calm assurance. The scientific revolution, which had been one of the fruits of the Renaissance, had continued into the times of Galileo and New-ton and had, by the end of the 19th century, reached its crowning cul-mination in the triumphant creation of the "new" science of classical physics. The "new" physics had been more than four centuries in the making.

All this time science had been the private concern of a few. Some-times a whole generation would pass between one discovery and the next. Only the boldest and luckiest of those "who not only departed from the opinions of the ancients but also correctly set themselves the aim of proceeding by the slow but sure path of experimentation" man-aged to achieve success or facilitate the success of others. "And they followed this path insofar as they were allowed by the short span of their lives or the multitude of their other affairs or the limits of their means." [3, p. 252] These are the words of Bishop Sprat, the author of "A History of the Royal Society" published in 1667.

In this book, Bishop Sprat for the first time elevates those who endeavour to discover the laws of nature to the ranks of the philosophers and scholars. The rejection of the physical description inherited from the ancient world had already been recognised as justified. Copernicus had already worked out the heliocentric system which had shaken the foun-dations of the religious world-view. Giordano Bruno had already paid with his life for daring to propagate the teachings of Copernicus, which the Church had condemned. (Giordano Bruno was born in 1548, five years after the death of Copernicus). Kepler, carrying on the observations begun by Tycho de Brahe, had discovered the laws governing the movements of the planets. Galileo Galilei had died a blind prisoner of the "Holy Inquisition" after being accused of heresy, and forced to yield to the Inquisition's demands at his shameful trial in 1633.

In 1643 Newton had been born. And yet the natural scientists were only now being recognised as the "third category of the new philosophers". Giving them their due, Bishop Sprat wrote that much had already been done and that "one may have doubt ... concerning only the attitude of future ages. And even then we can safely promise that they will not long be deprived of a whole galaxy of enquiring minds, for before them lies such a clearly marked path. They have only to taste these first fruits to be inspired by this example." [ibidem]

1687 saw the appearance of Newton's "Principia: the Mathematical Foundations of Natural Philosophy" which opened a new era in physics. There was no longer any doubt as to the verdict of "future ages". The interval between one major discovery and the next would no longer last as long as a whole human life but would be only a year or two. However, right up until the 18th century science would play only an indirect part in the development of the new industrial civilization.

Even then, only in the 19th century would science exchange its passive role for an active one, so that "the perfecting of cannons" ceased to be dependent only on "the skill of the foundrymen". Scientific and industrial progress were now closely intertwined and kept pace with the passing years.

The end of the 19th century saw the emergence of a new confidence, a feeling that dominion over nature was now complete. The telephone and the photograph were already invented. The steam turbine and the internal combustion engine were already at work. Soon radio would be born. The elegant laws which rested on the classical mechanics of Galileo and Newton, on Maxwell's electromagnetics, seemed unshakeable. The very air seemed to be permeated by ideas for wonderful inventions.

"On the horizon of classical physics there were two dark clouds, which darkened an otherwise clear sky: the work of Michelson and the problem of the distribution of energy of the black body radiation spectrum." [4, p. 143]

In 1879 the American newspapers announced the rising of a bright new star in the scientific sky. Naval sub-lieutenant Albert Michelson, who was not yet 27 years old, had accomplished an amazing feat in the field of optics: he had measured the speed of light. [5, p. 11] Michelson had not only achieved a very high level of precision in his measurements, he had also proved beyond question that the speed of light does not depend on the movement of the Earth. This affirmation directly contravened the laws of classical Newtonian mechanics, according to which the velocity of a light ray travelling in one direction away from the Earth must be greater than the velocity of a ray travelling in the opposite direction.

But neither the stunning result of Michelson's experiments nor the radiation spectrum of a heated body, which could not be explained on the basis of existing theories, worried the older physicists unduly. There was no denying the contradictions that had arisen, these "two black clouds" could not be ignored, but the physicists, secure in their established certainties, felt that, all the same, eventually the contradictions would somehow be reconciled. They felt that these minor difficulties could not stand in the way of the imminent completion of a description of the forces acting in nature, of the unified picture which was practically within their grasp. And so the scientists, still solitary figures working in small laboratories and "back rooms", went on plying their unhurried and enjoyable trade.

Suddenly, chaos broke out.

In November 1895 Wilhelm Konrad Röntgen, "at that time an obscure professor of physics in Würzburg, bought one of the new cathode-ray tubes, aiming to study its internal construction. Within a week he had encountered a mysterious phenomenon taking place outside the tube. Something was proceeding from it which had hitherto unimaginable properties – something which lit up flourescent screens in darkness and which exposed photographic plates even though they were wrapped in black paper. Furthermore, the photographs which were produced were amazing – they showed coins hidden in pockets and even the bones inside a hand.

Röntgen, having no idea what this "something" was, called it the "X-ray". The ray was there for all to see and it is not surprising that within a few days news of it had encircled the globe. It became the subject of innumerable music-hall jokes and within weeks all the most eminent physicists without exception had repeated the experiment for themselves and had demonstrated the phenomenon to a marvelling public." [3, p. 400]

Less than three months after Röntgen's discovery a professor at the École Polytechnique in Paris, a certain Antoine Henri Becquerel, heir to a fine collection of phosphorescent substances bequeathed by his father and grandfather, decided to check the cold rays produced by ordinary flourescent minerals and salts to see whether they possessed any such properties as those of Röntgen's rays.

Research into phosphorescence and luminescence was the Becquerel family's tradition. Antoine's grandfather Henri Antoine César Becquerel, an officer in France's regiment of engineers, had in later life become a professor of physics and a member of the Paris Academy of Sciences. He had devoted his long life to research into the phenomena of phosphorescence and luminescence. He discovered that certain substances were transparent to ultra-violet light and was the first to describe diamagnetic properties.

Antoine César's son, Aleksandre Edmond, who carried on his father's studies and indeed collaborated with him for many years (Edmond survived his father by only 12 years) worked out a scientific classification of the phenomena of phosphorescence and the basic laws relating to such phenomena.

Antoine Henri carried on the work of his sires. He was 44 when the news of Röntgen's amazing rays, which made it possible to "see your own bones", flashed round the world. Henri Becquerel selected a few samples from his huge collection of phosphorescent substances and exposed them to sunlight for a prolonged period. The result turned out to be negative: the bright rays given out by the samples after their thorough irradiation failed to penetrate matter.

Becquerel changed the samples. Among the new samples, which were chosen completely at random, were some thin laminae of uranium salts. These thin flakes left their image on a photographic plate protected by thick black paper. This was a success of a kind – but what did it mean? Instead of the uranium salts Becquerel might have taken any other substance and could, indeed, have gone through the whole collection without any kind of positive result.

At that moment, however, having seen the images on the plates, Becquerel did not know that he had hit upon a rarity. Much less did he suspect that he had discovered a previously unknown natural phenomenon. This phenomenon was to be confirmed by the Curies three years later. They used other elements, not phosphorescent substances at all, and called the phenomenon radioactivity.

Meanwhile, the experiment with the uranium laminae only encouraged Becquerel to hope that he would find what he was looking for – that is, some kind of resemblance between the rays emitted by the cold phosphorescent substances after a thorough preparatory irradiation in sunlight and the rays discovered by Röntgen.

However, yet another freak of chance convinced him that he had stumbled upon a mysterious phenomenon which was in no way connected with the Röntgen rays or other properties. While checking the result he had obtained with the uranium flakes, Becquerel selected a different sample of uranium salts and got it ready for its exposure to sunlight. However, for two days in a row the sky was overcast, the sun did not appear, and Becquerel packed away into his desk drawer all the objects he had prepared for the experiment – a photographic plate wrapped in thick black cloth, an aluminium sheet, a thin copper cross on top of that and, last of all, a piece of uranium salt.

A few days later he developed the photographic plate which had lain in the dark drawer and found on it the clear outline of the copper cross. He had shown that his uranium salts spontaneously, without any preparatory irradiation, emitted rays which were capable of penetrating a black cloth and an aluminium sheet.

Whence came this very considerable radiated energy? Whereas in the first experiment it was possible to assume that the sample absorbed the energy of the sun's rays and that this energy was then transformed into radiation, the second time no energy could have been absorbed but a chemically inert substance radiated energy – the origin of which was a total mystery. This contravened the law of the conservation of energy and was, therefore, inexplicable.

Becquerel tested other samples, too, but none of the various other substances which did not contain uranium gave any such result. On the other hand, all the uranium compounds, both phosphorescent and non-phosphorescent, whether in solid form or in solution, emitted the very same radiation. Its intensity depended solely on the proportion of uranium in the compound. It became clear that spontaneous radiation, independent of any processing of the sample, was a property of uranium alone.

In 1899, as a result of a painstaking chemical analysis of mica blende, Marie and Pierre Curie discovered two new radioactive elements, to which they gave the names polonium and radium. The Curies not only found that polonium and radium emitted Becquerel's mysterious rays. Another no less amazing property was also discovered: radium, without changing its appearance in any way, gave out enough heat to melt its own weight of ice. Why? There was no answer.

Rutherford was also at this time investigating radioactivity. In that same year, 1899, he established that there were three different types of radiation. One of them consisted of material particles flying at unimaginable speeds. Rutherford called them alpha particles and the radiation he called alpha rays. The radium atom, as it ejected these rays, transformed itself into an atom of a different substance – some kind of heavy inert gas – and an atom of helium. This was genuine alchemy.

Rutherford concluded that in radioactive elements spontaneous atomic transformations are taking place. Yet another basic law of classical physics was thus undermined – that is, the law of the immutability of the elements. The ancient alchemists, whose ideas had been so thoroughly demolished and ridiculed, could now have the last laugh – matter, without any human intervention, was transforming itself as the experimenters watched.

These discoveries marked the beginning of a series of bewildering developments in physics. They came in rapid succession, compressing events which might have been expected to take place over decades. Ruthlessly, each successive discovery pushed the previously sound logic of classical physics nearer to collapse. The "two little clouds on the horizon", which had been expected to disperse at any moment, had been joined by great thunder-clouds and now threatened imminent destruction of that elegant tracery of laws which had been erected with such toil and patience over four centuries. This was a disaster. The atmosphere of assurance and tranquillity changed to one of confusion and chaos. The 20th century had arrived.

There is a poem by Anna Akhmatova in which she sees the beginning of the new numerical century as the beginning of a period leading up to a revolution:

"The famed embankments saw its dawn –
No mere calendric change was this –
A genuine New Age was born."   [1]

The "not merely calendric" change in physics – the "New Age" – began two weeks before Christmas in the year 1900.

On the 14th of December 1900, at a meeting of the Berlin Physical Society, a professor at the University of Berlin, a certain Max Planck, read a paper in which he presented a solution to the problem of the energy radiation spectrum of a heated body. Planck blew away one of the "small clouds" only to raise much more fundamental questions. They were to be answered in a fantastically short time – and in the process modern science would be created. At that moment, however, neither Planck himself nor his distinguished audience had any idea that very soon, within their lifetimes, the hour when this paper was read would be recorded as the hour when modern physics was born.

For several years Max Planck had devoted his energies to the theoretical study of radiation given off by a totally black object. He had tried widely differing approaches to the problem but all had failed. Finally came the moment of which 20 years later at the Nobel Prize ceremony Planck would say: "After several weeks of the most intensive work I had ever done the darkness dissolved and new vistas, undreamed of hitherto, opened up before me." [4, p32]

Planck had at last arrived at a simple formula linking radiation energy with the frequency of emitted light but in order to achieve this he had been forced to abandon the laws of classical physics and introduce into science a completely new concept – the "action quantum" ("quantum" being Latin for "quantity").

Planck surmised that electrons radiate energy not evenly but in separate packets. Moreover, the energy is directly proportional to the frequency of the radiation and the coefficient of proportionality is always the same – i.e. is a universal value. He used the symbol h for this value and called it the elementary action quantum. Planck's constant would soon enter into all the formulae of the new physics, would remove all contradictions and restore order. However, at first even Planck himself, unable to trust the new constant since its physical meaning was still beyond his comprehension, described it only as "the formal symbol of a law successfully arrived at by guesswork."

In his Nobel Prize speech when the quantum was already 18 years old, Planck said: "The failure of all attempts to throw a bridge across the abyss which had opened up removed all doubts: either the action quantum was an imaginary value – in which case the whole formulation of the law of radiation was fundamentally illusory, a mere formulaic game devoid of real content – or the formulation was based on correct physical logic, in which case the quantum had a fundamental role to play in physics and its discovery heralded something new, hitherto unheard of, which would evidently require the transformation of the very bases of physics ... Experimentation decided in favour of the second alternative." [4, p. 39]

Actually, for the first five years there was no experimentation. In 1905 Albert Einstein's article "Concerning a Heuristic Approach to the Production and Transformation of Light" appeared in the journal "Annals of Physics". In the paper Einstein put forward the bold hypothesis that light consists of particles – light quanta.

In 1907 Einstein went even further. He published a paper in which he argued that any system exhibiting small fluctuations must contain an amount of energy which is a multiple of Planck's quantum – that is, the energy of any small fluctuation can be measured in terms of quanta. On the basis of this assertion Einstein was able to solve the problem of the thermal capacity of solid bodies. This was the problem which could justifiably have been called the "third cloud" on the horizon of physics.

Yet again, in 1909, it was Einstein who took the next important step in the development of quantum theory. This time he considered the theory of radiation fluctuations. Here was the first revelation of the dual nature of light – i.e. that it could be regarded as corpuscular or in terms of waves. The confidence with which Einstein used the quantum hypothesis did not please Planck in the least and, in fact, attracted his disapproval: "While many physicists for reasons of conservatism reject the ideas which I have formulated or are as yet uncommitted, others find it necessary to move on from my ideas to yet more radical positions of their own ... Since nothing is more harmful to the development of a new hypothesis than the application of the same beyond legitimate boundaries, I have always been in favour of tying in the quantum hypothesis as closely as possible with classical dynamics ... " [7 p. 7].

By describing his development of the quantum hypothesis as radical, Planck acknowledged Einstein's genius. In 1912 Europe's four leading physicists, Max Planck among them, nominated Einstein for membership of the Prussian Academy of Sciences and recommended that his ideas about light quanta, though they "went beyond legitimate boundaries", should not be held against him.

Einstein's scientific reputation at that time was founded on the success of his theory of relativity. Planck was one of the first to become interested in Einstein's relativity theory and, indeed, to understand it. In the 1905 "Annals", where Einstein's first paper on light quanta appeared, four other articles by him were also published. Each one of these five papers laid down guide-lines for the development of modern physics. Let us recall just two of them: "Towards an Electrodynamics of Moving Bodies" and "Does the Mass of a Body Depend on the Energy it Contains?"

The first of these is a full and clear exposition of the special theory of relativity. The relatively speedy acceptance of this work was evidently due to two important factors. Firstly, by this time much of the groundwork for the theory had been done; in Einstein's words, "there was no doubt in 1904 that the theory's time had come ... Lorentz (who had already laid the mathematical foundation for the relativity theory) now knew that the transformations later given his name were in agreement with Maxwell's equations and Poincaré had gone deeper into this correspondence." [8, p. 322] Secondly despite its revolutionary character, the theory of relativity did not contradict classical Newtonian mechanics, but retained it within itself as being valid for marginal cases involving velocities much smaller than that of light. It was this work which would bring Einstein his lasting fame.

The second article mentioned has no equal in the history of science. Consisting of only three printed pages, it contains the law of the equivalence of mass and energy – the key which opened the door to the use of the colossal energy released by nuclear reactions. The mathematical form of this law is extremely simple: $E=mc^2$. E is the energy, m the mass and c the speed of light.

(Every schoolchild knows this formula now. A few years ago I went to see a programme of short films. Some were sad, some were comic. One of these films was about a forlorn and diffident physicist, a reluctant bachelor. All the events in the film revolve around his unsuccessful attempts to get married. In one scene he is lecturing to his students. He writes a formula on the blackboard. In close-up we see the board and his hand writing first a capital E, then an equals sign, then a small m multiplied by c. Then, instead of squaring the c the hero writes a question mark: $E=mc^?$ A roar of good-natured laughter filled the cinema at this point.)

Immediately after completing his paper on mass and energy Einstein wrote in a letter to a friend: "From the principle of relativity, in conjunction with the fundamental Maxwell equations, it follows that the mass must be directly related to the energy contained in the body. In the case of radium there must be an observable loss of mass. This is a most gratifying advance in understanding." [9, p. 73] It was this very understanding, so "gratifying" at the time, which was to grieve Einstein at the end of his life.

"The tragedy of his (Einstein's) last years reached its peak with his intervention in the affair of the atomic bomb. Almost 40 years previously he had derived the formula $E=mc^2$ from the theory of relativity. Its significance was realised long before it was subjected to any kind of experimental test, much less embodied in any practical application. Now this possibility presented itself, along with the danger that Hitler might lay his hands on awesome weapons of mass destruction with which to hold the world in thrall. This prompted Einstein to write his famous letter to President Roosevelt, which spurred the atomic bomb programme and at the same time led to the present horrifying absurd situation, in which the human race is faced with the stark choice between peace and self-destruction," wrote Max Born in his recollections of Einstein [8, p. 398]. But this was all in the future. In the first years of this century scientists thought only of revealing the secrets of nature.

In the spring of 1910, Ernst Solvay was discussing the difficulties involved in the interpretation of the experimental data which were then accumulating with Walter Nernst, the professor of chemistry at the University of Berlin. Solvay was a chemist and wealthy industrialist who took a lively interest in the changing scientific scene. He suggested inviting Europe's leading physicists to Brussels to reflect together on the new situation in physics. He promised to pay travelling expenses and board for all who would participate in the conference.

Nernst got to work immediately. He wrote invitations to all the leading lights of European physics. The first letter was addressed to Max Planck. Planck's reply was not optimistic. He doubted whether the conference would arouse general enthusiasm and asked for it to be postponed for several years, until a time when more convincing results would be available. Nernst, however, managed to convince Planck, after which there was no need to persuade the rest. All those invited gladly agreed to come to Brussels.

The 30th October 1911 was appointed as the opening day of the conference, the theme of which was "The Theory of Radiation and the Quantum". Germany was represented by Nernst, Planck, Sommerfeld and Warburg; England by Rutherford; France by Brillouin, Madame Curie, Paul Langevin, Louis de Broglie and Poincaré; Austria by Einstein; Holland by Kamerlingh-Onnes; Denmark by Knudsen and others. Such was the beginning of the famous Solvay Congresses, every one of which proved to be a landmark in the history of science.

At the first Solvay Congress there was no clarification of the difficulties which had arisen. Looking back we can see that at that point no clarification was possible. However, the congress played an important part in the unfolding of events. Its main aim was achieved – i.e. the best minds of Europe were brought together and focussed on the problems. Science took a step towards becoming international.

Shortly after the congress Ernst Solvay founded the International Institute of Physics, to which he denoted a million Belgian frances. The Institute was to carry out scientific research, assist young researchers and convene the regular Solvay Congresses. The second such congress was arranged for October 1913. Its theme was the structure of matter.

1913 was a year of dramatic developments in science. It was the year when nuclear physics was born. Back in 1911 Rutherford had made his amazing discovery and had announced to his colleagues: "I know what the atom looks like!" For several years experiments had been going on in Rutherford's laboratory which involved the bombardment of various target substances with beams of alpha particles. Such particles penetrated thin plates with ease, except in those rare cases when one of them would for some reason be deflected. Rutherford suggested looking for these particles, which possibly were being deflected at a greater angle.

What a surprise for the experimenters when they detected particles which, on colliding with a thin metal sheet, returned in the direction whence they had come! Rutherford, once he was finally convinced that this was indeed happening, said; "This was the most unexpected event in my life. It is almost as improbable as if you were to fire a fifteen-inch shell at a sheet of cigarette paper and watch it bounce off the paper and come straight back at you." [10, p. 17]

On the basis of these experiments Rutherford put forward the idea that the atom is composed of a positively charged nucleus around which electrons orbit at great distances – great, that is, by comparison with the size of the nucleus. From the standpoint of classical physics such a model of the atom, resembling a miniature planetary system, seemed simply absurd. According to the laws of classical aerodynamics the electrons, moving in their orbits around the nucleus, should radiate energy and, in just 10^{-8} seconds, fall onto the nucleus. However, in the real world this does not happen. Experimental data, too, are hard to dismiss. What was going on?

In 1913 Niels Bohr, who was at that time working in Rutherford's laboratory in Manchester, published a three-part article. It was a trilogy which left the reader in "a state of shock". On the basis of Rutherford's model of the atom Bohr had created a theory of the atom – a theory which worked. Now the quanta had penetrated into the last refuge of classical physics – into the formulae of classical mechanics and electrodynamics.

Most impressive of all was the success of the theory in explaining the line spectra of hydrogen-like atoms. The first systematisation line spectra was the work of Johann Balmer and dates back to 1885. Even earlier, however, in 1870, Johnston Stoney had observed that the frequencies of the lines of the sun's spectrum, which corresponded to certain lines in the spectrum of hydrogen, were related to each other as whole numbers. He found a surprising analogy between the relationships of these numbers and the relationships between certain harmonics of a violin string. This had led Stoney to conjecture that underlying the laws of line spectra there must be some sort of periodic movement inside the hydrogen molecule.

Fifteen years passed before Balmer arrived at the general law linking the wave numbers of various lines of the visible spectrum to simple whole numbers. This law can be expressed very simply:

$$m = R\left(\frac{1}{2^2} - \frac{1}{n^2}\right).$$

Here m is the "wave number", the reciprocal of the wavelength; R is Rydberg's constant, thus named in honour of the Swedish spectroscopist Rydberg. Not even its existence could have been deduced from any known laws of physics, much less its numerical value, which is equal to 109.678. This value was found by a purely empirical method. It is constant, the number of a spectral line n assuming integer values, beginning with three. Balmer's law is highly precise, being accurate to five or six significant digits.

In 1904 Lyman found a series of hydrogen in the ultraviolet band which can be described with the use of the same formula, the only difference being that $\frac{1}{1^2}$ replaces $\frac{1}{2^2}$. In 1909 Paschen found a series of hydrogen in the infra-red area of the spectrum. The formula was again the same but the first term was now $\frac{1}{3^2}$. These whole-number laws of banded spectra, purely empirical and yet following the experimental data with such precision, seemed absolutely mysterious.

In the very first drafts of his theory Bohr arrived at a formula which described the laws of the spectrum. This formula contained as particular cases the series discovered by Balmer, Lyman and Paschen and predicted other series which were indeed discovered later. As for Rydberg's constant, in Bohr's formula this was a pure theoretical expression letter and consisted of Planck's constant, the charge of the electron, the mass of the electron and the velocity of light.

The conformity of Rydberg's constant as calculated according to Bohr's formula with its empirical value was staggering. How could anyone refuse to believe Bohr's theory now? In that same year, 1913, the validity of the theory was confirmed by the work of Franck and Hertz. The results of their experiment showed unambiguously that the internal energy of the atom cannot change smoothly or steplessly. It changes in very definite discrete steps.

This provided direct confirmation of Bohr's hypothesis concerning the behaviour of the electron. He had suggested that electrons follow very definite "permitted" orbits in the atom and that in so doing, in defiance of the laws of classical electrodynamics, they do not lose or radiate energy. Energy is radiated only when they change over from one "permitted" orbit to another. Moreover, the energy that is released is packaged in multiples of an indivisible "portion" which is proportional to Planck's quantum.

Of course, the triumph of Bohr's theory was the chief event of 1913 but there was also Rutherford's prediction of the proton and Henry Bragg's invention of the X-ray spectrometer. In that same hectic year Vladimir Konstantinovich Arkadyev, one of Lebedev's pupils, discovered the effect later called ferromagnetic resonance; Soddy introduced the term "isotope"; Aston proposed the separation of isotopes by the gas diffusion method; Kamerlingh-Onnes discovered the suppression of superconductivity in a strong magnetic field.

Physics was entering its Golden Age. The era of solitary geniuses had passed. Schools were forming: Rutherford in Manchester; Max Born in Göttingen; Sommerfeld in Munich; Ehrenfest in Leiden; Marie Curie in Paris. They all gathered talented young people around them. These were not, however, "different" schools or "personal" successes – they belonged to all.

Rutherford's young people, led by the Crocodile (Rutherford's nickname) himself, visited Ehrenfest's seminars. People came from Zurich, Berlin, and Paris. After the seminars the arguments continued in the street, in the cafe, in the study at Ehrenfest's home, that small room where every visitor left his autograph on the wall. In Göttingen, in Hilbert's famous Mathematics Club, physicists more and more often took the floor. They came from all the cities of Europe. That is how it was ... until it was brought to an end by a little word which rolled like thunder across the continent – the word "war".

War has no favourites; it demands all without exception. It has its own rights, its own aims. It pronounces its own sentence. It allows only one profession – that of soldier. Even war could not halt the growth of those new scientific ideas which were not directly related to military needs. Let us remember that it was at this moment that Einstein shaped the general theory of relativity. War, however, divided people into enemies and allies. The scientists, for the most part uninterested in politics, found themselves in the grip of political squabbles. The German scientists had the greatest difficulties to cope with. They belonged to the nation which had dragged mankind – 38 nations – into a bloodbath.

After the war communications between scientists of different countries, which had been blocked so completely, began to be restored. The German scientists, however, were excluded from all the international organisations and conferences. In 1919, after several meetings in London and Paris, the major European scientists founded the International Research Council (I.R.C.) [1,1]. Not one German was admitted. Memories of the war were still so fresh that many scientists were even in favour of excluding representatives of countries which had remained neutral.

The Solvay Committee set to work again (Lorentz was still its chairman). The next (that is, the third) congress was arranged for April 1921. The boycott of the Germans continued. No invitation was sent even to Nernst, Solvay's close friend and one of the main organisers of the first two congresses. Einstein was the one exception. The explanation given was that Einstein currently held a Swiss passport. However, Einstein refused the invitation, explaining that he was planning a trip to the United States.

In 1922 a group of eminent European scientists formed the International Union of Pure and Applied Physics with the aim of normalising relations between all the various countries – but all the founders' efforts to admit German, Austrian and Hungarian scientists to the Union ended in failure. The boycott went on.

The fourth Solvay Congress took place in 1924. The only physicist "of German extraction" to be invited was Einstein. Once again Einstein declined the invitation, this time stating openly that he considered that acceptance would amount to betrayal of his German colleagues. "In my view" – he wrote to Lorentz – "it is wrong to mix science with politics and also wrong to burden a person with responsibility for actions taken by the government of the country where he happens to live." [ibidem, p126]

It was a time when there was no suspicion that science could be harmful. The time when a ghastly threat would hang over mankind was still to come. Science was still sacred. However strong the prejudice against Germany, the desire to puzzle out the secrets of Nature proved overwhelming. Science became international once again. So began the second phase of what has come to be called the heroic period of modern physics.

From left to right: Einstein, Newton, de Broglie, Schrödinger,
Heisenberg, Dirac

Chapter 2

Fiat Lux

The very beginning of the 1920s was perhaps just as impressive as that magnificent year of 1925 from which quantum physics dates its anniversaries. The whole decade was a time when physicists were taking stock of the situation, when they asked themselves to what extent they really understood the experimental data that had been collected.

The experiments had produced alarming results. Only two things were certain: first, that the data definitely confirmed the reality of quanta; second, that the accepted methods of classical physics had proved totally inappropriate for the exploitation of the new knowledge. This was also the moment when it became clear that the quantum theories so far developed were in serious need of re-assessment. The world of physics was in such a white-hot state that when quantum mechanics appeared as if by magic it ripped through that world like an explosion.

From its birth, quantum mechanics took a variety of forms. Theories which seemed at first glance to be completely at variance were arrived at almost simultaneously by three different men working in three different cities – by Heisenberg in Göttingen, by Dirac in Cambridge and by Schrödinger in Zurich. Today these theories form the core of quantum mechanics and the foundation of modern physics.

However, for all this to come about great things had to be achieved. Heroic efforts were required of the older generation while the generation born at the same time as the quantum had to contribute an unbounded creativity. This latter generation had been born and had grown up in an atmosphere of constant change and discovery in all spheres of human activity. This generation's birth coincided with the first powered flights and the first radio transmissions. It was the time of the first inoculations against diphtheria, of the discovery of vitamins and of the identification of blood groups. It was the time of the discovery of peptides, the moment when the first hypothesis concerning the composition of proteins was being formulated. Explosives and fertilisers were being developed. Visitors to exhibitions were startled by a new kind of art.

In the midst of a tornado of change, accepting innovation as the everyday norm, grew up the generation which had been born with the century. It grew up in a world which had still not had time to get used to the first aeroplanes, a world which, while still remembering the Wright brothers' prophecy that "Mankind will not fly for another thousand years", had seen the growth of a mighty aeronautical industry. The air battles of the First World War had taken place soon after Lieutenant Nesterov had looped the first loop over Russia. Now inexpensive small cars scampered along the cobbled roads of Europe while Marconi's modest laboratory, which had sent the first wireless message across the Atlantic at the turn of the century, had grown into a money-making enterprise.

The new generation came into physics at a moment when great new hopes had given way to bitter disappointments – which were to be followed by further hopes and new disappointments. Only the experimental data were beyond dispute – and they were as stunning as they were unquestionable. As for their interpretation, the only thing that was obvious was that quanta had to play a fundamental role in physics. However, all the theories based on the idea of the quantum, though at first they seemed sound, were at best only in qualitative agreement and, as experimental data became available, were shown to be altogether unfounded. Physicists were obliged to introduce more and more limitations and additions into the theory, all the time struggling to make it fit the facts. Everybody realised that these adjustments were temporary measures and that there could be no consistent theory until fundamental changes were made. It was therefore a time of tireless searching for new ways forward.

Enormous efforts were made to find a way out of the difficulties and contradictions. Physics was like a tangled ball of wool from which several loose ends hung out. If skilfully tugged any one of these ends might at first begin to move – but would then only yank the central knot tighter. In the words of Niels Bohr, the whole situation in physics caused him to feel "sadness and hopelessness". [12]

The new generation, however, came to physics without such negative feelings. It had no time for the alarms and misgivings of its teachers. The younger men came to the science like knights, fearless and beyond reproach. They were bold and light of heart and, if they did fall into despair, then it was the despair of children, very deep but short-lived.

A detailed account of that brief period which preceded the birth of quantum mechanics – that is, the period from 1922 to 1925 – is the subject of a separate book. Here, just to touch upon that time, we shall recall the story of the discovery of spin. This is all the more appropriate as the main characters in this story – Pauli, Uhlenbeck, Goudsmit, Kronig – were born at the same time as the quantum or even later.

At the end of the preceding century Lorentz had predicted the splitting up of spectral lines in a magnetic field. In 1896 Pieter Zeeman confirmed this phenomenon experimentally. In 1897 Lorentz formulated his theory of the "Zeeman effect". In the following year Zeeman observed different, completely unexpected behaviour of spectral lines in a magnetic field. This was dubbed the "anomalous Zeeman effect". Lorentz's classical theory could not account for it.

All this was, however, so impressive that in 1902 Lorentz and Zeeman were awarded a Nobel prize. This was the second Nobel prize awarded for physics. Röntgen had received the first one in 1901. Concerning the Zeeman effect the next was to be that won by

Pauli in 1945 for work he had done 20 years previously – for his discovery of the principle which bears his name and which enabled him to explain the anomalous Zeeman effect.

Wolfgang Pauli (born 1901) studied under Sommerfeld. He was 19 years old when he first heard Einstein lecturing on relativity. Straight after the lecture Pauli stood up and said: "You know, what Herr Einstein has been telling us is not as stupid as it sounds."

In some book or other (I forget which, but the writer was not a physicist) I once read something to the effect that physicists are very brave, many of them being mountaineers, divers and skiers. The gist of what the author was saying was that if you cannot conquer Everest or ski straight down the Matterhorn then you will never make a real physicist. Pauli, however, was neither a climber nor a diver and by the time he was twenty he was already overweight and unusually clumsy – so clumsy, in fact, that in the laboratory people shrank from him.

No one who ever wrote about Pauli failed to insert some anecdote about his clumsiness. The classic example is the well-known story of the terrible explosion which took place in Göttingen, in the James Franck Institute. The cause of the explosion was a mystery. Then it was all explained quite simply. It turned out that at the moment of the explosion a train had stopped briefly on its way through Göttingen. Pauli was on it. The presence of Pauli for just a few minutes a mile from the Institute was quite enough to cause the disaster. The story of the meeting of the famous Ehrenfest and the young Pauli is revealing. After the first few minutes of their conversation the gentle, well-mannered Ehrenfest could not help remarking that he found Pauli's published articles more agreeable than their author. To this Pauli immediately replied that in relation to Ehrenfest he felt the exact opposite.

And so, towards the end of 1924 Pauli explained the anomalous Zeeman effect, first formulating the so-called exclusion principle, which had no parallel in classical physics and which seemed completely unfathomable. According to this principle an atom cannot possess two or more electrons which are at the same energy level.

In the theory which had been accepted up to this time the state of the electron was defined by three "quantum numbers" which corresponded to its three physical characteristics: energy, orbital momentum and the projection of the momentum in the direction of the magnetic field. All three values were expressible as quantum numbers: n, l, and m. To these three quantum numbers Pauli added a mysterious fourth number which described "the peculiar ambivalence, which classical physics could not describe, of the quantum-mechanical properties of the radiating electron." [13]

The "radiating" or "optical" electron is the last electron on the outer shell of the atom. In monovalent substances (hydrogen and the alkali metals) a single electron is to be found in the outer shell. For all other elements it is precisely the number of these outer, optical electrons – or their absence – which determines the valency of the element (they are thus arranged in the columns of Mendeleev's table).

By this time the physicists had come to the definite conclusion that in the case of the monovalent metals the angular momentum of the "atomic residue" – that is, of the nucleus and of the neutral envelope excluding the outer electron – is equal to $^1/_2$. Pauli, however, not taking this fact into consideration and speaking obscurely of the "peculiar ambivalence, which classical physics could not describe, ... of the electron ..." took another important step by declaring that the angular momentum of the atomic residue is determined only by the outer electrons.

It seemed obvious, in view of all this, that the electron possesses its own momentum equal to $^1/_2$. This in turn explains the halved momentum of the atomic residue. Secondly, it confirms the reality of the fourth quantum number, which corresponds to this same momentum of the electron. Finally, it explains the "exclusion principle". Perhaps this last statement should be clarified.

The fact is that in Bohr's model of the atom two or more electrons with the same quantum numbers can be present in the same shell. How, then, can Pauli's principle work, essential though it is for the understanding of the anomalous Zeeman effect? Here, too, it is clear that it is sufficient to recognise that the electron has its own momentum and that the fourth quantum number corresponds to it. Then there is no reason why two electrons cannot have identical quantum numbers – except for the last: their momentum of $1/_2$ must simply be orientated in opposite directions. This same $1/_2$ momentum of the electron – which is called spin in recognition of the fact that the electron rotates around its own axis – truly is a property of the electron and no phenomenon at the atomic level can be explained without reference to it. But it was not Pauli who introduced the concept of spin. In this connection bewilderment is expressed by all who remember Pauli. He was so close to spin and yet he missed it! But Pauli did not simply miss spin.

There is more to the dramatic story of spin than the fact that Pauli failed to grasp it. In a letter to the famous spectroscopist Landé Pauli expounded his own views concerning the exclusion principle; it amounted to a preview of an article he was preparing. Pauli's letter arrived at the same moment as a young visitor from America – Ralf de Kronig.

Kronig recalls: "On 7 January 1925, when I was just 20 years old and very inexperienced, I arrived in the picturesque little German university town of Tübingen and checked into the "Golden Bull" hotel. I had come to represent the University of Columbia at a meeting with Landé and Gerlach, who were respectively heads of the departments of theoretical and experimental physics.

Landé welcomed me warmly at the Institute of Physics, observing that I had come just at the right moment as Pauli was due to arrive the next day. In fact Pauli had written him a long and interesting letter, which Landé showed to me. ... . Pauli stood so high in my estimation that, as I waited for his arrival, I eagerly plunged into the letter which Landé gave me to read. It was a clear and critical exposition of the exclusion principle, written in characteristic style. ... . Pauli's letter impressed me hugely and naturally I wanted to grasp the implications of the fact that each separate electron must be describable in terms of quantum numbers already known from the spectra of the atoms of alkali metals, in particular the two momentum quantum numbers l and s ($=^1/_2$). Evidently we could no longer ascribe s to the outer shell and the thought flashed through my mind that s could be considered to be the momentum quantum number of the electron. In the language of the models that were the only available framework for discussion before the advent of quantum mechanics, the electron's own momentum could only be visualised as its rotation on its own axis.

This idea leads to a number of difficulties. However, it is an attractive image and, under the influence of the letter I had read, I arrived by the evening of that same day at a formula for the so-called 'relativistic doublets.' " [14, p. 15] The next day Pauli came and an argument ensued. Kronig received a complete rebuff. In his usual scornful fashion Pauli told Kronig that his idea about the spin of the electron was pure nonsense, that a mathematical point could not rotate – only a physical entity could do that. Soon afterwards Kronig left for Copenhagen where he told Bohr and Heisenberg about his idea. The Copenhagen school also rejected his rotating electron. The verdict of these eminent physicists had its effect and Kronig did not publish his formula.

However, in the autumn of that same year Ehrenfest's young colleagues Uhlenbeck (born 1900) and Goudsmit (born 1902) came independently to the idea of the rotating electron. They also used the term "spin" for the momentum of the electron. Uhlenbeck wrote: "Goudsmit and I arrived at this idea after studying Pauli's article in which he formulates the exclusion principle and for the first time ascribes four quantum numbers to the electron .... We felt that this so lacked foundation and caution that there must be some kind of mistake in it somewhere. After all Bohr, Heisenberg and Pauli himself, our chief authorities, had never before suggested anything of the kind.

We told Ehrenfest all about it .... Goudsmit and I felt that perhaps it might be better not to publish our ideas for the time being but when we told Ehrenfest of our intention he replied: 'I sent your letter to the publishers some time ago – you are both young enough to be forgiven for a silly mistake.' " [14, p. 246]

Thus it was that in the annals of physics Uhlenbeck and Goudsmit are recorded as the discoverers of spin – but such is the irony of fate that now, when all this is history, the concept of spin is inseparably linked with the name of Pauli. By the end of 1925 Niels Bohr was convinced about spin. Let us note that this did not come about without some assistance from Einstein. Heisenberg, too, accepted spin, taking his lead from Bohr. Pauli, however, would not budge. He sent dire warnings to Bohr and expressed severe disapprobation of his apostasy, his fall into scientific heresy. Pauli's wrath, however, was fairly quickly assuaged; it was impossible to deny the reality of spin. How frustrating everything was can be judged from Pauli's letter to Kronig in May 1925 (matrix mechanics was to arrive on the scene in June): "Physics is once again stuck in a blind alley, at least as far as I am concerned. It is just too difficult. I would rather be a cinema comedian or something of the sort and never hear about physics again!" [14, p. 34]

And yet so much had been accomplished! De Broglie's work had been published by this time. At the beginning of the 20s one of the most acute difficulties was that presented by the problem of wave/particle dualism. This was not a new problem. Back in 1905 Einstein, making free, in Planck's opinion, with the quantum hypothesis, had explained the photoelectric effect by ascribing corpuscular properties to light. Although Einstein's paper enjoyed deserved success physicists did not think of abandoning the wave-like nature of light. If light was not a wave, neither diffraction nor interference phenomena could be accounted for. As for the explanation of the photoelectric effect and the assumption of the corpuscular nature of light which it involved, each physicist understood it in his own way. For example, some considered that Einstein's light quanta were not particles but some sort of measure of the energy of an electromagnetic field.

In 1922 the American physicist Arthur Compton discovered the effect which is named after him. While investigating the scattering of X-rays in paraffin, Compton noticed that apart from the expected effect – that is, radiation with the same wavelength as the incident radiation – there was also radiation with a greater wavelength. Moreover, it proved to be the case that the shorter the wavelength of the incident radiation the smaller was the proportion of the scattered radiation which retained its original wavelength. If radiation is described purely in terms of waves then there is no explanation for this phenomenon – the wavelength of the scattered radiation should remain the same as the wavelength of the incident radiation. In other words, light should not change colour when it is reflected. This is in fact the case as long as we are dealing with wavelengths such as those of visible light.

Compton, however, was studying the scattering of X-rays and gamma rays, the wavelengths of which are tens and hundreds of thousands of times shorter than the wavelengths of visible light. He found out that in the case of hard gamma rays (which have the very shortest wavelengths, of the order of 10^{-12} cm) none of the dispersed radiation preserved its original wavelength.

A year later Compton (and also Debye, working independently) had arrived at a theoretical explanation of this phenomenon based on the premise that light was purely corpuscular in nature. This account adheres strictly to the classical laws of conservation relating to the deflection of two particles as they strike each other – as strictly, in fact, as would an account of the collision of two billiard balls. The only unusual circumstance was that one of the "balls" – the electron – was a normal particle (at least, it was so considered at the time) while the other one was a light particle the energy of which was not defined as its mass multiplied by the square of its velocity but as Planck's constant multiplied by its frequency.

The discovery of the Compton effect, then, meant that a solution to the puzzle over the dual nature of light was even more sorely needed. The time had come when, in Bragg's words, physicists were obliged to believe on Mondays, Wednesdays and Fridays that light consisted of particles while on the remaining days they had to believe that it consisted of waves.

Also in 1923, while they were coming to terms with Compton's experimental results and his theoretical explanation thereof, the bewildered physicists were hit by another piece of work which, being purely theoretical, was even more unnerving. Louis de Broglie, an enthusiastic amateur who spent his time in experimental work on radiation in a private laboratory owned by his brother, published three theoretical papers in quick succession. In these he asserted not only that light possessed a dual nature but also that matter itself, while remaining matter, must also consist of waves. That is, every material particle, as well as possessing physical properties such as mass, dimensions etc. must also be its own corresponding wave. So electrons and atoms and even molecules can all be regarded as waves. De Broglie defined the wavelength of any given particle as Planck's constant divided by the momentum. Therefore, if a billiard ball is regarded as a wave then its wavelength at a speed of, say, one metre per second appears as the first non-zero figure twenty-five places after the decimal point, In other words, the wavelength of the billiard ball is of the order of 10^{-25} cm.

With unusual lucidity De Broglie explained why it was necessary to regard the electron as having the properties of a wave: "The discovery of fixed movements of electrons in atoms requires the introduction of whole numbers but hitherto the only phenomenon in physics which have had descriptions involving whole numbers have been the phenomena of interference and natural oscillation." [15, p. 398]

Nobody took De Broglie's articles seriously. Whenever he tried to expound his ideas at a seminar they would cause general merriment in the auditorium. The first person to pay serious attention to them was Einstein. In 1925 he wrote to Born about one of the articles: "Read it! Even though it gives the impression that it was written by a madman it is a solid piece of work." [15, p. 399] But Born did not bother to look into it and his young associates even organized a spoof seminar satirizing De Broglie's ideas.

In the same year Einstein brought these same ideas to the attention of Erwin Schrödinger. Schrödinger was 38 years old at the time but it was not only his age which set him apart from the younger founders of modern physics, from the alumni of the dynamic "quantum schools". Schrödinger was a loner. He never belonged to any of the schools. He was born and brought up in Vienna where, as well as taking courses in philosophy and history, he also studied physics and mathematics. Under the powerful influence of Hasenöhrl, who had taken over the chair on the death of the great Boltzmann, he fell in love with these subjects. Schrödinger's main passion in physics were the statistical methods which originated in the pioneering work of Boltzmann.

Passion is the right word in Schrödinger's case. He was infatuated with biology (he wrote on the evolution of the human eye), languages, poetry (there is a small volume of his verse), sculpture (his room looked more like a sculptor's studio than an academic's study) and, above all, philosophy. "I was slow in coming to the modern theory of the atom." – wrote Schrödinger – "Its internal contradictions are like ringing dissonances compared with the pure, inexorably clear consistency of Boltzmann's thought. There was a moment when I was practically ready to flee. However, I was urged on by Exner and Kohlrausch and found my salvation in the study of colours." [16, p. 39]

It was his work in visible spectroscopy, to which he brought a profound knowledge of the theory of vibrations, as well as other "external" factors which led Schrödinger to wave mechanics. As regards the "external" factors the most important, perhaps, were the following: firstly, a close association with Hermann Weyl, who at that time headed the mathematics department at the university of Zurich and who helped Schrödinger to develop his mathematical competence; secondly, the friendship of Peter Debye who had explained the Compton effect in 1922; finally, the "punch on the nose" as Schrödinger put it [17, p. 331] which Einstein gave him for his "vain attempt to construct a wave model of an electron in an elliptical orbit".

While delivering this "punch" Einstein also pointed out the importance of De Broglie's work. De Broglie's ideas, mad though they were and conceived by a man who was not even a professional physicist but an art historian, fascinated Schrödinger. This was the time which Schrödinger's wife remembered as the most troubled period of all. Her husband, normally so placid and such a good family man, lost his peace of mind and started frightening her by saying that he had made a discovery which rivalled that made by Newton. This discovery – Schrödinger's wave mechanics – was to be revealed in 1926.

In 1929 De Broglie was to be awarded a Nobel Prize. Meanwhile in 1925, while Schrödinger was just getting started on his work, Heisenberg created the first version of quantum mechanics – matrix mechanics. Heisenberg was in his twenty-fourth year. His task as Born's assistant was to decipher the same old atomic spectra. The semi-classical foundation on which Bohr's theory was built proved unsteady. The interpretation of the spectroscopic data went no further than the qualitative level.

Even in the simplest case – that of the hydrogen atom – the interpretation would not bear rigourous scrutiny. All Heisenberg's attempts to base his work on Bohr's model of the atom proved unsuccessful. He decided therefore to abandon this model and, indeed, to free himself from all assumptions as to what happens inside the atom. Instead, following a "hunch", he took as his starting point the empirical data, the external indications of the movement of the electrons – that is, the actual spectral data. This new approach led to the long-sought breakthrough; it enabled him to develop an adequate mathematical framework for the description of the electrons in the atom.

It is worth repeating that Heisenberg took only the actual empirical data as his point of departure. The match between calculation and actual result proved to be excellent. Heisenberg realized that he was on the right track but one striking fact caused him to doubt. The symbols with which he operated behaved strangely. If he took the product of the momentum of the electron p and its coordinates q, then the difference between qp and pq was not equal to zero. Instead, this difference was invariably equal to Planck's constant multiplied by the square root of minus one. Could it be that the "hunch" and the elegant theory based on it would turn out to be illusory?

Heisenberg worked on this problem on the island of Heligoland in the North Sea, having fled there to escape the hayfever season in flowering Göttingen. Though not very happy with his algebra, Heisenberg nevertheless wrote his article and took it to show Max Born.

"...he was a very talented assistant", wrote Born, "but still very young and inexperienced. He did not even know precisely what a matrix was. He got into difficulties and came to me for help. After some thought I found the link between his ideas and the matrix calculation. I remember how surprised I was to discover that Heisenberg's quantum condition was in fact the matrix equation $qp - pq = ih$. I then worked out the matrix version of quantum mechanics in collaboration with my pupil Jordan." [8, p. 237]

Born immediately sent Heisenberg's article to the publishers. Heisenberg went to work with Bohr in Copenhagen but took part in the development of the theory. "There was a very lively correspondence", wrote Born, "but my letters were unfortunately all lost because of some political disturbances. The result was an article signed by three authors which more or less completed the formal part of the research. Before this article had time to appear, however, a dramatic development took place; Paul Dirac published an article on the very same topic. Stimulated by a lecture that Heisenberg had given in Cambridge he had arrived at conclusions similar to ours in Göttingen – the only difference being that he had not borrowed the matrix theory from the mathematicians but had independently discovered and elaborated the principle of non-commutation." [8. p.306]

Paul Dirac (born 1902) produced his version of quantum mechanics at the age of twenty-three, supporting it with a mathematical framework of his own creation. At thirty he occupied the chair which had once been Newton's. In 1928 he was to formulate the equation which would unite the basis of the theory of relativity with quantum mechanics and would mark the beginning of quantum field theory. Of this equation Dirac himself remarked with some pride that it explained the major part of physics and the whole of chemistry.

So it was that in Cambridge yet another version of quantum mechanics was born – Dirac's version. Max Born wrote: "While we were discussing all this, yet another dramatic event took place – Schrödinger's famous articles appeared. His thinking had been developing on completely different lines and could be traced back to Louis de Broglie." [8, p. 306]

The scientific world was thrown into confusion. Two kinds of matrix mechanics had already been formulated in Göttingen and Cambridge and here was yet another kind of mechanics which was diametrically opposed to the first two and yet just as beautifully consonant with the evidence. In the summer of that same year 1926 Heisenberg went to Munich to visit his parents. He chanced to attend a meeting addressed by Schrödinger.

"This was my first acquaintance with the interpretation which Schrödinger wanted to put on his mathematical dualism – i.e. wave mechanics. I fell into despair at the thought of the conceptual muddle which, I was sure, would be introduced into atomic theory as a result of this interpretation. Unfortunately nothing came of my attempt at clarification in the ensuing debate. I argued that if we accepted Schrödinger's view it would become impossible to explain Planck's law of radiation but no one was convinced. Wilhelm Wien, professor of experimental physics at the University of Munich, answered me quite sharply saying that indeed there would now be an end to this quantum leap and atomic physics in general ..." [17, p. 375]

It was not long, however, before Schrödinger himself showed that there was no contradiction whatsoever between wave mechanics and Heisenberg's matrix mechanics and they were reconciled under the name "quantum mechanics" – a term which Max Born had first introduced into physics in 1924.

We are indebted to Born for something even more important; it was Born who revealed the meaning of Schrödinger's equation as well as the uniquely correct meaning of the famous wave function which Schrödinger had introduced as a function describing a wave in its "pure state" (a premise which turned out to be absolutely unjustified). In Max Born's interpretation the wave function is not only not a wave but in fact has no physical reality at all. The only thing which has any meaning is the square of the wave function which determines the probability of finding a particle in a particular energy state at a particular point in space and time.

To the end of his days Schrödinger could not accept that Born's interpretation in terms of probability was the correct one – but it is the very essence of quantum mechanics. It was precisely with this interpretation that Schrödinger's equation turned out to be the fundamental equation of sub-atomic reality, the equation upon which modern physics is built. Today we know that everything is much deeper. Schrödinger's equation rules not only the sub-atomic world – it also describes a wide variety of phenomena in the macroscopic realm.

There will be no detailed account here of the long storm which raged around quantum mechanics after its birth – or of the famous dispute between Einstein and Bohr. The great Einstein, to whom more than to any other of the great men we owe our present knowledge, remained until the end of his life unconverted to the "quantum faith".

Werner Heisenberg was awarded a Nobel Prize in 1932 for his pioneering work in quantum mechanics. Erwin Schrödinger and Paul Dirac received theirs in 1933. Such was the creation of quantum mechanics. Now the time had come to apply it.

"At the end of the 1920s the echoes of that great explosion of knowledge, the creation of the theory of relativity and of quantum mechanics, had rolled to the furthest corners of the world," wrote Yuri Borisovich Rumer in his recollections of meetings with Einstein, "A multitude of young people widely differing as regards talent and level of preparation rushed to the centres of the new 'quantum faith' – to Copenhagen, Göttingen, Zurich, Leiden, Cambridge – in order to share in this feast of the imagination and the intellect. I was twenty-eight years old when I turned up in Göttingen in the summer of 1929." [18, p. 108]

Gänseliesel

Chapter 3

To Göttingen

Legend has it that in 1737 George August, Elector of Hanover, commanded two of his cities, Celle and Göttingen, to build a university or a prison. Each town was to state which of the alternatives it preferred. The citizens of Celle clearly had more money and more sense than those of Göttingen: after some reflection they decided to build a top quality prison. After all, nothing stands so long or brings such benefits to such a variety of governments and regimes as a good jail. Nothing else serves so well in the cause of justice and the education of youth as unfailing retribution in a tangible form. Students, on the other hand, lower the moral tone with their impudent wit and disturb the peace and prosperity of a town with their singing and dancing. In this way the good people of Celle arrived at their decision, certain that they had chosen wisely – but they were wrong. Rather few people are aware even of the existence of the town of Celle. Göttingen, however, which chose to build a university, has won world-wide fame.

It was the will of fate that George August's university should be a meeting place of great minds: the king of mathematicians, Karl Gauss, lived and worked here from his student days until his death; Weber and Riemann worked here; the Grimm brothers taught Germanic philology; here Groutefend made a start on the decipherment of ancient Babylonian cuneiform script. Anyone who visits Göttingen's cemetery is bound to be struck by the number of world-famous names.

By the beginning of the 20th century Göttingen had become, in the words of Max Born, the world's "mathematical Mecca, the glory of which rested on the teachings of three prophets: Felix Klein, David Hilbert and Hermann Minkowski". Then, as soon as theoretical physics made its appearance, the world of physics also came to know the names of Göttingen scholars.

On the whole it was just a normal, quiet, almost sleepy little provincial town, like many others in Germany, and had not its inhabitants shown such foresight 250 years ago, perhaps no one living more than 200 miles away would have heard of it.

I have in my possession a little old album of photographs of Göttingen published before the First World War. In style and appearance it is just like so many others of its kind. In it there is a description of the town: "Situated in the southern part of the territory of Hanover in the wide hollow of the upper valley of the river Leine (at 150 metres above sea level), Göttingen nestles picturesquely at the western foot of Mount Hain, whence magnificent woods and parkland roll down to the edge of the town. To the east and north, Mount Hain merges into the majestic wooded heights of the endlessly rolling Göttingen limestone plateau (300–423 metres above sea level) with its romantic declivities.

Having sprung up near the castle Kaiserpfalz Grona, the town was a dukedom by 1387. In the medieval period strong fortifications were built as is testified by the defensive ditch which is still to be seen, now adorned with limes. Since 1737 this has been a university town, freed of its surrounding ramparts in 1764. A district court sits in Göttingen and there is a railway station, the town being situated on the Hanover – Kassel and Göttingen – Bebra – Frankfurt railway lines. The station, which is still as originally built, can no longer cope with the growing traffic and is due for reconstruction." [19, p. 2]

On a warm July day in 1929, on a train from Berlin, Yuri Borisovich Rumer arrived in Göttingen. Leaving his fibre suitcase in left luggage, he went straight from the station to Max Born's Institute of Theoretical Physics. The whole Institute turned out to consist of the professor himself, his three assistants and a Frau something-or-other who kept the place tidy and did the typing. Another of her duties was to get Professor Born's car ready each morning.

Finding Max Born's study without any difficulty, Rumer opened the door and saw the professor. An American was explaining his ideas to him in broken German. Turning to the latest visitor Born said: "Just sit down, I'll be with you in a minute" and went on listening to the American. He showed extraordinary patience and benevolent attentiveness but could not make head or tail of the exposition. Finally he sat back in his chair and said: "Tell you what – try to write all this down in English and I'll read it."

Having seen the American out he turned to Rumer:

"Have you got an idea, too?"
"Oh, yes, Herr Professor, I've got an idea!"
(Rumer was too modest to add that his idea was unique.)
"Uhum" – grunted the professor – "Go on then, let's hear it."

The young man from Russia started to talk. He had Moscow University, with its traditionally strong school of mathematics, behind him. He had brought with him a paper in which his idea was expressed in elegant mathematical form. With all the self-assurance of youth he explained it , longing only for one thing – that it should be valued as it deserved to be. Suddenly he heard Born's voice: "Mensch! You speak German well but I'm afraid all this is of no interest to me."

The young man came to a dumbfounded halt. However, the young stand firm and are not easy to deflect from their purposes. He carried on confidently as if he had not heard Born's remark. A few minutes passed and then Born asked a question. Rumer answered it. Born's sceptical expression changed to one of interest. There were more questions. The conversation went on and on. Now it was clear that Born was enjoying the company of this young Russian. Not only did he have a brilliant grasp of mathematics; he also had no difficulty in expressing his ideas in Born's native language – he even joked in German. Eventually Born, evidently satisfied, got up, walked up and down the room and said: "All right. Maybe I'll take a chance on you."

Soon Born would write a letter to Einstein: "A young Russian turned up here recently. He has a six-dimensional theory of relativity ... At first I was very doubtful about him but he spoke very intelligently and soon convinced me that there is something in his ideas. Since I am less of an expert than "ε" in these matters I have sent his work to the Göttingen Academy. I shall send a copy to you, too. I urge you to read it and assess it. The young man's name is Rumer. If his work impresses you favourably I would like to ask you to help him. He knows all the mathematical literature from Riemann's geometry to the very latest publications and could make you an ideal assistant. He is a personable young man and gives the impression that he has been extremely well educated. His address is: Georg Rumer, Oldenburg, Festungsgraben 8." [1, p. 101] The letter is dated 12 August 1929.

After that first meeting with Rumer, Born called his two assistants Heitler and Nordheim and asked them to help Rumer to settle in. It turned out that the three were all of the same age. They were soon friends. At the beginning, of course, they sounded each other out. For the first half-hour they talked about science, then the conversation moved on to various other topics. In carrying out the task which Born had entrusted to them, Heitler and Nordheim walked from one end of the town to the other, trying to find Rumer somewhere to live. Along the way the two young men fell over each other to describe the town to the young Russian. Here were the best bookshops in Europe, wonderful concerts, Bach, Haydn – and incredible cakes. Basically the town consisted of students, small shopkeepers, landladies and Nobel prize-winners.

For a long time Göttingen had remained unchanged. Contented with their lives, the burghers of this quiet little town had been in no hurry to try the unsettling innovations of the twentieth century. They looked at their town and its surrounding countryside. They liked their own homes and the little wine cellars where the beer, said the two guides, was excellent. This feeling of contentment was everywhere. It was there in the good-humoured way they spoke to foreigners who knew no German but were striving to penetrate the mysteries of this surprising town. It was somehow expressed by the way they drank in the beer cellars and even in the amused glances they cast at the one or two automobiles that had made their appearance. It was as if they were quite sure that these monstrous products of a sick mind would never stand a chance against real, live horses. "You know", said Nordheim, "Horses have a special status in Göttingen. Not just in Göttingen, but in the whole of Hanover. You know how people think that the English are terrible ones for their traditions - medieval soldiers in fur hats and all that? Well, the English love of tradition is nothing to the love of tradition you will find in Germany. Most of the local burghers consider themselves Old Hanoverians, quite independent of the Prussians and the Saxons and the rest. And the horse is their symbol. You'll see this horse everywhere – in the parks, on the front of buildings. Out in the country there's always a little statue of the horse of Hanover even on the humblest dwelling. It was this symbol that the Prussians once trampled on. In 1866 the Prussian troops came, deposed the Kurfurst and took over the whole state of Hanover. Ten kilometres from Göttingen there is an estate where you can still see an old notice: "By the authority vested in me under the temporary regulations promulgated by the commander-in-chief of the Prussian army of occupation in February 1866, I order all horses to be removed ..." The fact is that these temporary regulations of 1866 are still in force to this day, since no-one has ever rescinded them. The local people are still resentful and regard the Prussians as aggressors.

Before they knew it, the young men were in the town centre: "This is Wilhelmplatz. In the guidebook it is called the town's 'jewel box'." Wilhelmplatz did not resemble a square at all. It looked more like an inner courtyard with beautifully arranged flower-beds. It was enclosed on all sides by severe Romanesque buildings, the main jewel in the box being the Aula – the great congregation hall of the university. In this building, apart from the assembly hall itself and the council chamber where the Göttingen Academy of Sciences meets there is also a punishment cell where those who broke the rule forbidding duelling served their sentence. It is said that the smell of the iodine which the students used on their rapier wounds still lingers in the cell. Do not imagine that duelling is purely a thing of the past. Real purists still fight with rapiers.

Until recently the Aula, as well as being used for assemblies, was also the only large lecture hall that the university had. The way the timetable was put together was another unchanging tradition. Because there was such demand for the Aule, the older professors had priority – that seemed only right and proper in the old days. The young lecturers often found it difficult to gain access to any lecture room at all. When the timetable was being fixed, people would come from all the faculties. The old men, the more eminent ones, would send their assistants (if they had them) as their authorised representatives. On a big blackboard squares were marked off as if for a chess tournament. The days and hours were listed along the horizontal axis while vertically, there was a list of the names of the professors in order of seniority. It was not his subject that counted, nor the ability of the lecturer – the only thing that mattered was how long he had worked in the university. If he had been there the longest then he was the first to state which courses he wished to teach and to fix the times for his lectures. No one, least of all the professor doing the choosing, was required to check whether one course clashed with another. There were times in Göttingen when electrodynamics was not taught for five years in a row, simply because of timetable clashes.

The student, too, was free to choose which lectures to go to. In Göttingen, as in the other German universities, students were not expected to attend a whole course of lectures and then be examined on it. Everything depended on what you intended to do afterwards. There were only two kinds of examination: doctors' examinations and assessors' examinations. Why so? The reason was that no one could enter government service without the title of assessor. Anyone hoping to become a burgomaster or a bank manager or to hold any civil service post had to pass the assessors' exam.

The assessors' exam, however, could not be taken at Göttingen. The university considered that just to have studied at Göttingen was qualification enough. In order to take the assessors' exam students had to travel 200 km to Celle, famous for its hard-labour penitentiary which served the whole country. There they had to supply brief details to the effect that they had been in Göttingen, had attended the lectures of such and such a professor and taken part in such and such seminars. No check whatever was made as to the accuracy of this information. The candidate simply had to produce a piece of paper requesting permission to take the exam.

The Celle examination syllabus was a little more demanding than that of the schools but well below the level required at university. The level of difficulty also varied somewhat according to the background of the candidate; law students, for example, were always given easier examinations. As a group, law students were considered weak candidates as the majority of them were destined to be assessors rather than doctors.

The doctors' examinations were a different matter altogether. Anyone wishing to graduate as a doctor had to wait until he was offered a chance to be examined – that is, until his teachers thought him worthy of becoming a doctor of the University of Göttingen. By comparison, obtaining a doctorate in mathematics at, say, Heidelberg was a much simpler and faster process. At Göttingen you had to wait patiently. It was possible to give up and leave but you were not allowed to ask to be entered for the doctors' examination. Only when the professorial body had considered the case carefully and come to the conclusion that the young man in question would not ruin their reputation in the academic world, was an examination arranged. There could only be one of three results: cum laude (with distinction); eximie (excellent); egregie (outstanding). Then the scholar received his doctors' diploma. It was printed on white paper. The name of the new doctor was also printed. The diploma was rolled up tightly and placed in an attractive cardboard tube. Along with the actual diploma the scholar received five extra copies – also printed. These would be needed later as proof of status when the graduate applied for jobs.

At the end of this long procedure the newly-honoured doctor was required to walk to the market place not far from the Aula. There, surrounded by 16th-century timber-framed buildings with the Rathaus on one side looking like a fortress, is a delicate open-work fountain with a statue representing a little goose-girl – the famous Gänseliesel. The young doctor, after being rewarded his degree, is required to climb up on this fountain and kiss the goose-girl. Since the University of Göttingen was founded no one has ever broken this tradition. The citizens maintain, quite correctly, that of all the girls in the world this little maid has been given the most kisses.

Thus Göttingen lived its slow and regular life. It was proud of "its sausage and its university" and could never forgive Heinrich Heine, who had given the girls of the town a reputation for having fat legs. Life went on without major changes, apart from advances in science, until the First World War. In the war years the lecture rooms and the library were as empty as the bierkellers. The great David Hilbert was regarded as a traitor because he would not put his name to a declaration made by some of the most eminent people in German academic and artistic life in which they assured the world that they were fully behind the Kaiser. For one whose birthplace was East Prussia, where the spirit of the Teutonic Knights lived on, this was considered unforgivable.

After the war some returned with Iron Crosses, some did not return at all. To the old duelling scars, which had been worn so proudly in the old days, far deeper and more grievous wounds were added. For some, the war had been a nightmare the like of which must never be seen again. For others, defeat was a disgrace which must be expunged with blood.

When peace came Europe was showered with American gold. The Americans felt that science was becoming a good investment. In Göttingen, just as in other major centres of scientific research, emissaries arrived representing the Rockerfellers and the Duponts who cared nothing for any boycott and did not think in terms of enemies versus allies. All over the town around Georg August's old university new institutes sprang up. There were new laboratories full of the very latest equipment, new buildings with large, medium and small lecture rooms, all founded on transatlantic profligacy. There was even a luxury hotel on the square in front of the railway station, most of its rooms as yet unoccupied. Streets were torn apart as little old houses were demolished and replaced by new ones – still small but with central heating this time.

Altogether there was plenty to say and plenty to show to the young man from Russia who had just arrived in Göttingen. The time had flown so quickly that the new friends had not noticed that it was already dark. They had to find a room quickly. On Plankestrasse Nordheim stopped in front of a pink-painted, timber-framed house and said: "I'll go in. You wait here. I'll find out if they've got anything." A few moments later he returned. "Come on. The landlord's name is Müller. He loves meeting young people from abroad. If you tell him that you're going to be corresponding with Einstein he'll pay for the honour of having you as a lodger."

The young men went in and were greeted by a man with a black moustache. He looked like a Spaniard. It turned out that he had a room to let with its own separate entrance. When the foreigners descended on Göttingen a large number of such rooms became available. The young men quickly came to an agreement with the landlord and, enormously pleased with each other, said goodnight. No less satisfied with his new acquaintance, the landlord offered good money on the spot, in advance even, for any letter that Einstein might send. He offered the young Russian a cup of tea – or would he like coffee?

As they drank their tea Herr Müller told Rumer what a fine town Göttingen was, how peace-loving were its citizens, how delicious its sausages. He gave advice about the best places to buy sausage and cheese and where to eat well and cheaply. Finally, as businessmen often do, he started bemoaning the state of his own affairs: "I am by profession a trader in ostrich feathers. At one time I used to make a good living from them. I was quite rich. Now they've gone out of fashion. Ladies don't wear them any more. And I can't seem to stick to any other trade. My comrades and I (Rumer was surprised to hear him use the word "comrades") have done our best to revive the fashion. We went to Paris. We talked to the very best designers and tried to get them to bring this lovely old fashion back. You can't beat ostrich feathers. We even offered them quite a large bribe. But no matter what we offered, no matter how obvious it was that we were right about the potential profits, they were not interested. Ladies just don't buy them, they said. That's why I'm reduced to messing about in small-time business for the rest of my life."

It was a long time before Rumer got to sleep that night. The events of the day and all the things he had heard kept spinning round in his head – and stopping at the most striking fact of all: that none of his new acquaintances had asked him about his own amazing country. No-one had asked what life was like in Russia or what the Revolution was all about. Yuri Rumer, the Moscow lad from the Maroseika, could have told them a thing or two.

"Konka" in Moscow

Chapter 4

From the Maroseika to Chistiye Prudy

Yuri Borisovich Rumer was the youngest child of the Moscow business-man Boris Yefimovich Rumer and his wife Anna Yurevna Brik. Their other sons Osip and Isidor, born a year apart, were respectively 18 and 17 years older than Yuri. His sister Lisa was 11 years older.

Apart from the love and affection of their parents the Rumer children also enjoyed the love and affection of their governess Alisa Bleker. This lady came from one of those German immigrant families who had lived in Moscow since time immemorial. She loved all the children equally and devoted herself equally to the education of each one. She brought them up to speak German like natives and would read to them until she was hoarse from Schiller and Goethe. When the Rumers were away on holiday on the Black Sea or abroad she used to stay in their home and look after the household. She would take no payment for any of this, being herself well provided for. She never married and used to say that she loved Boris Yefimovich but had, rather than break up his family, dedicated herself to the education of his children.

The Rumers occupied the whole first floor of the Yegorovs' apartment house on Kosmodemyansky Pereulok (SS. Cosmos and Damian Alley), just off the Maroseika. The alley had got its name from the Church of SS Cosmos and Damian which had been rebuilt in the 19th century by the Russian architect Kazakov on the corner of the Maroseika and what is now Starosadsky Pereulok ("Old Garden Lane"). Restored and repainted, it stands there to this day.

The Maroseika, the central of the three main thoroughfares of the eastern part of the "White Town" began immediately after the Ilyinsky

Gates at the edge of the Kitai-gorod. After the Kremlin and the Kitai-gorod, the White City was the third concentric ring of urban growth in the history of Moscow's expansion. In the reign of the last of the Ryurikovich, Tsar Fyodor Ivanovich, in the 1590s, a white stone wall was built around it. 28 towers and 10 gateways were incorporated in this semi-circle of wall, which by the end of the 18th century had fallen into disrepair. The empress Catherine II decreed that it should be demolished and that boulevards should be laid out along its former course.

These eastern areas of the White Town were occupied partly by merchants but mainly by craftsmen. Preserving a tradition from those early times, the Solyanka area remained the centre of the salt trade, storing salt and salt fish for the whole of Moscow in its cellars. Directly adjoining the Maroseika was the Grand-ducal part containing a suburban residence and stables used by the Tsars. All the Grand Dukes and Tsars of Greater Russia, right up to Peter the Great, rode from there along the Maroseika to the Kremlin.

When the wealthy aristocracy of the province of Moscow began to build town houses with large courtyards on the Maroseika, the traders and craftsmen began to move out. The other newcomers who pressed into this area were the "Germans", a name applied to all foreigners in those days. They bought up the land, paying two and three times the usual price. Alongside the palaces of the gentry grew up a large "German" suburb consisting of a maze of narrow lanes, stretching from the Maroseika to Chistiye Prudy ("Clear Pools"). Packed along these lanes were rows of neat little German houses. The Germans built "ropaty" (little halls for Lutheran prayer meetings) and worked mostly in skilled trades. They were excellent apothecaries, confectioners, and tailors.

In the 17th century the proximity of these "infidels" to the Kremlin and the Kitai-gorod along with the fall in the number of Orthodox households in the parish prompted Tsar Mikhail Fyodorovich, founder of the Romanov dynasty, to move the Germans from the Maroseika to the Gorokhovoe Polye (the "Pea Field") on the river Yauza. Their new settlement was three versts (approx 3 km) from the

Pokrovsky ("Intercession") Gates. The Germans soon built new homes and before long another bustling, polyglot suburb had sprung up.

Along the Maroseika meanwhile, and along the little lanes leading off from it, the gentry built themselves new residences, one of these being the home of the famous boyar Matveev. It was here that Tsar Aleksei Mikhailovich first met Matveev's ward, Natalya Naryshkina, who was soon to become Tsaritsa and the mother of Peter the Great. Also on the Maroseika lived the Miloslavskys, the boyar family from which Tsar Aleksei Mikhailovich's first wife had come. On Starosadsky Pereulok was the palace of the hetman Ivan Mazepa and, indeed, a whole Little Russian colony which served as the Moscow base for official representatives from the Ukraine. In fact the name "Maroseika" is derived from the name of this "Malorossiisky" (Little Russian) colony. In quick, everyday use this adjective was eroded to "Marossisky", losing the "lo".

Along both sides of the Maroseika there were shops and the dwellings of the "streltsy", the professional officers drawn from the nobility. The streltsy paid a tax in kind by having soldiers billeted on them. However, the presence of common soldiery in a nobleman's house was not particularly welcome and therefore special huts with their own street entrances were often built in the forecourts. In the gloomy reign of Pavel (Paul) the First, who increased this tax in kind, the distressed gentlefolk petitioned the Tsar to allow them, at their own expense, to build barracks for the soldiers outside the city. Pavel agreed and the burdensome regiments were evicted from the Maroseika.

In the time of Peter the Great, who abolished the "Pale of Settlement" for foreigners, the "Germans" again began to buy up property along the Maroseika. Peter also allowed them to build their own churches. Once again a "German suburb" grew up along the Maroseika. The little lanes and alleys began to disappear. Peter banned the old practice whereby each house was built in the middle of its plot and commanded that houses should be "built along the line of the street and roofed with iron". The first multi-storey buildings, intended for commerce or bureaucracy, now sprang up.

On the Maroseika there also appeared apartment buildings, not particularly dignified as regards their architectural style. They were intended as rented accommodation for the better-off. It was in one of these buildings that Boris Yefimovich Rumer, merchant of the first guild, lived. On the third floor of the same building lived the famous Moscow lawyer Kogan who shook the courts with his eloquence. The lawyer had two daughters, the future Lilya Brik and Elza Triole. In time, the Rumers and the Kogans became related. The elder sister, Lilya, married Yuri's cousin, Osip Brik who was Anna Yurevna's nephew and a frequent guest. The younger daughter, Elza, although she had been born four and a half years before Yuri, was his faithful childhood playmate. She had even taught him how to kiss when he was seven years old.

The land straight opposite their house belonged to the German Lutheran church. Apart from the church building itself there was a gymnasium where community meetings were held. When cinematography came along, films were shown in this same hall. There were also well-equipped tennis courts and other sports facilities. Along-side the Lutheran Church was the Lutheran school where every year there was a big party with a Christmas tree and everyone who went was given sweets. All the children on the Maroseika, irrespective of their religious affiliation, were invited to the Christmas party. These children learned their languages as they are learned in a bilingual family. No teachers were needed, of course. The youngsters just grew up speaking both languages – they would start a sentence in Russian and carry on in German or start in German and finish in Russian.

To this little corner of Moscow the Germans brought an atmosphere of tranquil efficiency not untinged, naturally, with a certain Teutonic pedantry and stiffness. The people walked with that typical German bearing and measured pace. You might think that a regular officer in civilian clothes was walking along the street when in fact it was only an art teacher. But it was plain to see that he was pleased to be an art teacher, pleased with his pupils, his home and his country – for that was how he regarded Russia. These particular Germans had no Fatherland but Russia. The German consul might take most of the young men off to do their military service in Germany and they might come back glad to have seen a little of the world – but they were also glad to be home. Some of them brought German girls back with them and in such cases, as a rule, would take them straight from the consulate to their church for the wedding ceremony. This did not happen often, however, because it was considered that these girls from Germany had not had the best upbringing and were not prepared for the life they would have to lead in Moscow.

Goethe somewhere recalls: "When I was 21 years old I loved Luisa above all other girls." In the collected works a footnote has been inserted: "At this point Goethe is mistaken; careful textual analysis reveals that at the age of 21 he, in fact, loved Charlotte above all others." A person who has grown up in a German environment will not be surprised by such a remark; accuracy comes first!

People who live in close contact and who are similar in character often come to be friends. That is probably how it was in the case of Boris Yefimovich Rumer and a certain pharmacist, Ferdinand Bleker. Bleker was a bachelor and probably quite wealthy. He was a partner in the "Verein" pharmaceutical company, a trace of which remains to this day in the form of "Chemist's Shop No. 1" on what used to be Nikolskaya Ulitsa (Nicholas Street) and is now 25th October Street. Ferdinand Bleker rarely went anywhere apart from his own private laboratory, where he spent his time searching for cheaper and quicker ways to produce medicines.

The Blekers and the Kogan family were often invited to the Rumers' home. While the children played, the adults would spend the long evening discussing the unsettling things which had arrived with the 20th century – instability in politics and wondrous advances in technology. Just recently they had arranged a special evening excursion to the Lubyanka to see the electric lights which had replaced the gas lights on that ancient square. They were still marvelling at the speed with which the electric tram had taken over from the horse-drawn variety. At this rate, they speculated with some apprehension, they would soon have to use electricity in the home, too, but they doubted whether such an innovation could be of much benefit or last very long. They had better hang on to the paraffin lamps, just in case. Indeed, although eventually domestic electricity came to stay and replaced the paraffin lamp, nevertheless in many households the old appliances were preserved in working order. The telephone had begun to appear in the wealthier homes. In those days it was easy to work out who had their telephone put in before whom: from the fact that the Kogans' number was 3-18 while the Rumers' was 7-15, it was clear that the Kogans had decided to have the telephone installed several months earlier than the Rumers.

Ferdinand Bleker and Boris Yefimovich often made music together. Bleker would play the piano and Boris Yefimovich the violin. They always played the same things: either Mozart or Bach. They either could not or did not wish to play music by any other composer. The programme changed only when Kogan joined them to form a trio. Ferdinand Bleker got on well with children and used the opportunity to show that he regarded them as grown-up people. However, he did not wish to be addressed by his name and patronymic in the Russian manner and remained "Onkel Ferdinand".

Onkel Ferdinand had four sisters. They were all very well educated and very pretty. The eldest, Maria, married a German baron from the Baltic provinces. Before the First World War they moved into Russia proper and took Russian nationality. Maria died before the Second World War, in which her sons were killed. The second sister, Mathilda, was the headmistress of a girls' grammar school and then some sort of government official. After the Brest-Litovsk treaty in 1917 she was sent to Germany. Then came Elen about whom little is remembered apart from the fact that she was unusually beautiful. She married a pharmacist and went with him to Nikolsk-Ussuriisk in the Far East before the war with Japan. No more was heard of her thereafter. "Liebe Alisa", who dedicated her life to Boris Yefimovich's family, became a member of its inner council, taking part in all discussions and decisions. When the time came for the two elder boys to go to school their father was inclined to give them a commercial education. Alisa insisted that they should go to a "Gymnasium" (grammar school) where they would learn the classical languages. Boris Yefimovich was somewhat perturbed by this, as the entry of Jewish children into the grammar schools was controlled by a quota system. Before they could take the entrance examinations, such children had to be interviewed by a commission upon which sat a high-ranking official from the ministry of education. Moreover, each year's quota was fixed, not according to the number of available places, but according to the number of applicants.

The first to get through the interview was the elder brother, Osip. His performance at the interview was so brilliant that he was exempted from the entrance examination and given a grammar school place without further formalities. The same thing happened the following year. This time it was Isidor, who was 11 months younger than Osip, who was offered a place at the Armenian school. The two brothers could not have been less alike. Sensitive and shy, ready at any moment to go off and shut himself in his room, Isidor impressed everyone he ever met with the profundity of his knowledge and of his thinking. Isidor had few amusements but, whatever he took an interest in, he soon knew inside out. For example, he had a wonderful knowledge of mathematics which he studied by himself, reading the works of Gauss, Riemann and Lagrange in the originals. It was he who translated the first popularising book on the theory of relativity into Russian. He made his living with literary translations. He was impetuous and unstable. At one point he announced that he intended to become an Orthodox Christian and take up a military career. However, the main passion of his life was philosophy. Isidor was a true philosopher. Unfortunately for him, he was an idealist philosopher, a disciple of Teichmüller. This was subsequently to be his undoing. Before his abrupt and traceless disappearance Isidor Rumer served as a personal secretary to Trozky.

While the younger brother lived a life of storms and passions, that of Osip Borisovich was the exact opposite. From his youth he was well known for his translations. He went on to become an outstanding linguist. He knew 26 languages including Latin, Greek, all the Western European languages and the languages of the East. He was a brilliantly gifted translator of poetry, he had a deep knowledge of history and culture. He wrote: "It seems to me that in many ways the role of the translator in literature resembles that of the performer in music." Osip Borisovich died in Moscow in his 72nd year, a luminary in the literary universe, leaving behind him a whole constellation of translators who had been taught the principle "Do not be afraid to depart from the letter of the text if by so doing you more nearly approach its spirit and meaning." [20 p3]

Boris Yefimovich, being a merchant himself, was not entirely happy with the dubious professions chosen by his elder sons and hoped that his younger son would go into a more profitable line of business. He decided that if "properly guided" the boy might make a good engineer. He was encouraged in this hope by Yura's precocious mathematical ability and clear distaste for ancient languages.

The classical "gymnasium" was not the most appropriate school for the boy, given this aim. The heavy weighting of the curriculum in favour of the ancient tongues and the almost complete neglect of the natural sciences meant that young people leaving such schools were thrust out like blind kittens into the new technical age. It was normal in the classical grammar school to devote 24 teaching periods per week to Russian language and literature, 34 periods to Latin, 24 to Greek, 22 to mathematics (including arithmetic in the lower school) and 6 periods to physics – i.e. less than a fifth of the time spent on Latin.

When Yuri was 10 years old, therefore, it was decided that he should go to a technical school. Although matriculation from such a school did not confer the right to enter university, it did open the way into the higher technical colleges. Just in case, Yuri's elder brothers prepared him for grammar school entrance and taught him Latin bit by bit. Their method was to open the grammar book and order him to learn by heart two pages of rules (with examples) before the evening. That was exactly the method that was used in the grammar schools. Yuri Borisovich always considered himself to have only a smattering of Latin – no doubt because of the way he learned it. He went on to learn many languages in the course of his life, including Persian and Hungarian, and usually was able to pick them up easily. However, because of this lack of Latin he considered that he had failed to live up to the family's reputation, which dated back to the old days on Kosmodemyansky Pereulok – i.e. that "all the Rumers speak all languages."

At that particular moment there were also other reasons for preferring the technical school. In 1910 Kasso became Minister of Education. Under his direction the schools were run like detention centres, the most repressive measures being reserved for the classical "gymnasia". One of Kasso's special edits, for example, forbade grammar school pupils from appearing in the streets in the evenings, even under parental supervision. All entertainments were banned during Lent. Any innocent prank was treated as a crime, even on occasions as a political crime. The headteachers of grammar schools were required to submit a report on each pupil, giving details of behaviour throughout the school year. Here is an example from the archives: "... 24th April. Late for the lesson following lunch break (the miscreant being a 13-year-old grammar school pupil – Author) and when rebuked by the Head invented absurd excuses. By order of the council confined to the punishment room for 6 hours. Parents notified. In addition, on five Sundays, after attending the Liturgy, to be detained for three hours." [21, p. 122]

The severe regime in the grammar schools influenced Yuri's parents in their choice of a school as much as his dislike of dead languages. Unlike his drilled and disciplined elder brothers, the carefree and incorrigible Yuri was up to all kinds of mischief. During that last summer of liberty before his first term at the technical school, Yuri played a trick for which his father did not find it easy to forgive him.

Yuri and his mates were mad keen on stamp collecting. The stamp dealer on Mylnikov Pereulok was the object both of their worship and of their envy. The things he had were amazing; everything from old English stamps to the latest issues from Northern Borneo. The treasures they saw in that shop seemed to them the most precious that the world could offer.

However, stamp-collecting came to a sad end for Yuri when his father forbade it. This is how it happened. One day, Yuri was in the stamp dealer's shop. He happened to see an envelope covered in interesting stamps and with the address of Thomas Cook's Cairo office in one corner. Yuri copied the address and wrote a letter to Cook's in his father's name. He gave them to understand that he was a wealthy Russian businessman wishing to make a journey from Cairo to Khartoum and from there by camel to Djibouti (his father and mother had, incidentally, just got back from Italy). This letter apparently caused some excitement in Cairo, for a reply came very quickly by the standards of the day. Boris Yefimovich received a letter from Cook's in one of the firm's own envelopes. It was covered with exotic stamps. Cook's wrote that they had been delighted to hear from such a wealthy and distinguished client and that everything, of course, would be arranged. As for the camels, Mr Rumer was assured that they were already reserved.

Yuri was the technical school's top pupil. He was presented with all the prizes and certificates that any pupil could be presented with. Even his pranks were forgiven. In all his time at the school he received only one serious punishment (detention every day for a week) and that was for a capital offence.

It was in 1913 – a year packed with momentous events. Strikes had become commonplace. The Tsar was celebrating the tercentenary of Romanov rule. History lessons were devoted to the glorification of the Tsar and all his dynasty. Yuri managed to distinguish himself in one of these lessons.

Uncle Ferdinand had a parrot, of which the Blekers were extremely proud. The parrot was very large and of some kind of special breed. Even its feathers were much smoother than those of other parrots. It could only say two things: "Vive la Republique!" and "Vive la Revolution!" The Blekers proudly maintained that the parrot was very old and had lived through the French Revolution, only they were not sure which one – that of 1848 or the Great Revolution of 1789. Somehow Yuri smuggled this parrot into a history lesson and got it to say its piece. Yuri was not the only one who was punished. All the boys in the class, who had applauded the parrot and had enthusiastically joined in when it shouted its slogans, also spent a week in detention.

Events soon began to move with such bewildering rapidity that these boys, who had hardly left behind the carefree days of childhood were forced to grow up in a hurry. On the 27th July 1914 the Tsarist government began a general mobilisation and Germany declared war on Russia on the 1st August.

Many years afterward, when Hitler came to power and a new menace hung over Europe, Osip Borisovich was to say: "How wrong people were to think that the First World War started, ran its course and ended. Mankind has still not paid the price for allowing that war to happen."

For a time, however, in August 1914 when hordes of refugees were pouring out of the forward areas, nothing much changed on Kosmodemyansky Pereulok. Neither the Russians nor the Germans realised at first the gravity of the situation. They did not know that their old contented comfortable life was about to change radically. When, during the first days of the war, the Irish lady came as usual to give Yuri his English lesson Anna Yurevna was surprised to see her: "Good gracious, there's a war on. It isn't the time for lessons. When it's all over, next month or the month after, we'll start again." But the war turned out to be more savage than they thought.

Alisa Bleker, who had been staying with her eldest sister Maria not far from what is now Palanga, managed, despite her German passport to get back to Moscow quite easily. She came to see the Rumers at once. Anna Yurevna, kissing her as always, said: "Alisa, my darling, just look what your Germans have done!" "It's terrible," replied Alisa, "who would have expected the Germans to behave like this?"

Moscow was in the grip of a patriotic frenzy which was whipped up mainly by members of the "Union of the Russian People". Solemn litanies were sung, beseeching the Lord to "preserve the health of our Sovereign the Emperor". Processions carried portraits of the Tsar, ikons and banners. There were attacks on Germans.

Yuri Borisovich remembered how he and his friends went to watch these unheard-of spectacles. On the Petrovka was a shop – "Müllers Grand Pianos". Müller sold various musical instruments but mainly concert pianos. These very grand pianos were dragged out on to the pavement. They were not easy to smash, however, so the rioters handed their torches and portraits of the Tsar to the children for safe keeping and set about their work breaking the strings with their boots and smashing the shop window with the thick piano legs. They also smashed the chandeliers and everything else they could lay their hands on, then scattered the piano keys, which were covered with a thin veneer of ivory, all along the street.

Also on the Petrovka was a shop owned by a certain Einem, who sold chocolate. Einem's son was in the same class as Yuri. This shop was attacked during one of the anti-German riots and the local dogs were licking the pavement and the road surface for a week afterwards.

Not far from Kosmodemyansky Pereulok was a haberdashery who made his living mainly by selling white collars. He hung up a big poster which read: "The proprietor of this shop is Andrei Levinson – a Russian of the Jewish faith." He stood grandly behind his counter, secure in the knowledge that this time somebody else's shop window would be smashed.

Soon came the decree from the Tsar that all German citizens were to be removed from the capital and other cities to special settlement areas – i.e. they were to be interned. Boris Yefimovich had contacts in the police force and he did all that was required – that is, he obtained the necessary document from the chief of police, stating that at the request of Mr B.Y. Rumer, Merchant of the First Guild, Miss Alisa Bleker was granted permission to remain in Moscow. However, the war had an unfortunate effect on Lisa. She was at first offended and then angry when people spoke of the horrors of the war in her presence: "How can the

Germans terrify anyone? The Germans are gentlemen ...”; “You attacked us, but we’ll show you ...” etc. In short, Alisa Bleker, despite her life-long connection with Russia, turned into a rabid German nationalist. Yuri Borisovich was to see the final result of this unexpected transformation just over twenty years later when he visited her in Berlin. By then she was grey-haired and shrivelled. She stroked his head, very happy to see him, but at the same time did not consider it inappropriate to repeat revolting slogans concerning the extermination of the Jews.

Alisa decided not to stay in Moscow after all and went off to a place near Yekaterinburg with Onkel Ferdinand. They left from the Yaroslavl station. The whole Rumer family went to see them off. The escorting soldiers were quite friendly and the officer in charge took the trouble to look through his list and take them personally to the right carriage. It was a first-class carriage, too. At first the occasion was in no way different from so many other departures for the Black Sea or on foreign trips. It was only when the lieutenant said: “Would the ladies and gentlemen who are leaving take their places, please!” that the women started to cry. Onkel Ferdinand kissed Anna Yurevna’s hand. Then he gave Yuri a man-to-man handshake, saying in Russian, though he had always spoken German to Yuri before: “To me you have always seemed a decent sort of fellow. Try to stay like that.” The Blekers were not allowed to write from their place of exile.

The war went on. By the end of 1916 the superior strength of the Entente was obvious. Germany found herself on the brink of a political and economic catastrophe and felt compelled to offer to negotiate with the Allies with a view to ending the war. However, the Entente’s high command rejected this offer and went over to the offensive on almost all fronts. The revolutionary movement was growing in Europe, strikes were becoming widespread, there was a mutiny in the French army and mass anti-war demonstrations began. Russia exploded in the February Revolution and the Tsarist autocracy was gone for ever. Germany’s offer of a separate peace treaty to the Provisional Government which replaced the Tsar was met with a rebuff. The war still went on. October 1917 was approaching.

Clockwise: Mayakovsky, Khlebnikov, Eisenstein, Luzin

Chapter 5

Petrograd – and Back to Moscow

For Boris Yefimovich Rumer the end of Tsarist rule in the February Revolution was almost a foregone conclusion. He had little interest in politics but, as an educated and progressive man, he believed that this change was for the better and was prepared for it. What saddened him was that the absurd and cruel war dragged on. The family's material circumstances were much reduced. Lloyds, the company which Boris Yefimovich dealt with, would no longer insure vessels in the war zone and ships were very difficult to charter. For this reason Boris Yefimovich was obliged to move to Petrograd.

The elder sons, who were not involved in their father's business, stayed in Moscow. Yuri was now in the 6th class with one more year to study before leaving the technical school. This fact could have complicated the move but in fact things worked out in a most satisfactory manner.

After the Revolution, many people who had been exiled by the Tsarist Government began to return. There were many well-educated people among them, some of whom had been exiled along with their families. Special intensive courses were arranged for the children of these families, the purpose being to help them to take the examinations that they had missed and to catch up with their contemporaries. Boris Yefimovich found out about these courses quite by accident and suggested to Yuri that he should take the examinations for the seventh class as an external candidate and get his matriculation certificate a year early. The aim was to arrange for Yuri to be admitted to university at the age of 16.

Boris Yefimovich was sure that a university course would be a good foundation for an engineering career and had in mind the faculty of mechanical mathematics, not doubting for a moment that the "mechanical" part of the name was the better guide as to nature of the courses offered. This was just the opportunity that Yuri needed. He passed all the seventh-class examinations with top grades and he was awarded a first-class matriculation certificate. In August 1917, by special permission of the Rector of the University of Petrograd (the word "Imperial" had now been dropped from its title) Yuri was allowed to take the entrance examinations. Having passed these without difficulty he was registered as a student in the faculty of mechanical mathematics.

Today's University of Leningrad is a cheerful, crowded, noisy place. It is difficult to imagine the unheated corridors and the empty rooms of 1917 when there were no regular classes and lectures were given only on odd occasions. Nevertheless, students were students even then. They gave their own vitality to their alma mater and were nourished in return. They would turn up and wander the corridors, hoping to find some kind of lecture or discussion group. More often than not they would gravitate to the library.

The treasures of the University library were famous. The young people were able to dig out not only books on physics and mathematics but also excellent translations of Hafiz, Saadi and Omar Khayam in French and German.

Meanwhile, in the streets outside, the storm of October 1917 had burst. "I once heard Kollontai address a meeting and once I heard Lenin. That's all – but I remember it well", recalled Yuri Borisovich.

In the first days of his student life Yuri struck up a friendship with another under-age scholar, Boris Venkov. Venkov was fascinating company. He was completely at home in number theory, which seemed like a kind of magic to Yuri. He knew the history of science and of literature and was deeply interested in the Orient. They first met at the main entrance of the University. Yuri saw a lop-eared boy, resembling

Yuri himself, but with his hair shaved off. The boy sat on the steps, writing out formulae.

"Excuse me, are you a mathematician?" Yuri asked, most respectfully.

"Yes, I am. Mathematician Boris Venkov at your service." An imp of mischief nudged Yuri into asking:

"Have you long been a student of this art?"

"It is impossible to study this art, as you put it. A man is either born a mathematician or will never be one".

"Truly?!"

"If you have come to mathematics from mathematics, you are fortunate. If, however, you betray her and take up other sciences, the gods will be wrathful and when it suits them they will punish you".

Yuri was delighted. From that moment and for the next eight long months the boys were inseparable. Thereafter fate was to toss them in different directions – but for the moment they shared the joys of mathematics in the University's freezing lecture-rooms. Was there any topic they did not read or argue about? There was even a time when the great professor Uspensky lectured just to the pair of them ... Seeing them in the corridor one day, Uspensky went up to them and asked; "Are you bored, gentlemen?" Then, not bothering to establish which faculty they were from or whether they were in fact bored, he added: "You know, my depression is getting the better of me. Come on, let me give you a series of lectures on issues in non-Euclidean geometry." Not only did he give these lectures to the two boys, he lectured as if to a packed auditorium.

On one occasion the youths were waiting for the professor in "their" lecture room and looking through a book on Riemann's geometry. The door opened and instead of the professor a handsome young sailor walked in. He looked as if he had just stepped out of a film: he had gingery hair and a moustache with upturned ends; he had wise-looking eyes – and a machine-gun slung over his shoulder. He spoke very slowly and calmly: "You kids, make yourselves scarce or you might get shot by mistake. Just get lost for a while. We've got a job to finish off". The boys could not make out what this job was – revolution or execution?

"Excuse me, how long should we stay away?"
"Just while we get things sorted out".
"And how long will that be?"
"I don't know. As long as it takes. It's a serious matter".
"I'm hungry," said Yuri suddenly, surprising himself with his own words. Then he felt ashamed.
"Now you're talking like a man," said the sailor and turned to a companion: "Vanechka, run and tell them to send up some grub. We'll feed this pair – and our lads must be starving, too".

This was late afternoon on the 24th of October (Old Style) 1917.

Boris Yefimovich Rumer did not waste too much time agonising over what course to take after the October Revolution. He could not go along completely with the Bolshevik way of looking at things but he made a firm decision to stay where he belonged and to serve the new order faithfully and honestly. His friends and fellow-businessmen, those who, like him, had decided not to leave the country, elected him as their spokesman and entrusted him with the job of coming to an agreement with the Soviets of Workers and Soldiers' Deputies concerning the re-allocation of the stores and property which they had in their possession. Boris Yefimovich went straight to Krasin. It is necessary to understand what kind of man Krasin was. He was a rich businessman, a talented engineer – but his main career was revolution. In 1900 Krasin had finished an engineering course at Kharkov Technological Institute and was sent to Baku to work in the oil industry. He became one of the founders of the revolutionary "Iskra" network in the Caucasus. "I have the feeling that I ought to arrest Krasin", said the chief of police in Baku, " but I haven't the heart to arrest such a fine gentleman." [22, p. 7]

Krasin commanded the respect of all kinds of people in all kinds of circumstances. Perhaps it might be permissible to recall one vivid illustration of this – Krasin's meeting with General Ludendorf. At the beginning of 1918, things were going so badly for Russia that the main question was how to get out of the war and conclude some kind of peace. However, it was proving impossible to establish a dialogue with the enemy, especially after Trotsky's call for "neither war, nor peace". Negotiations were finally broken off on the 18th of February and the Germans launched an offensive along practically the whole Eastern front. The man chosen for the immensely difficult job of re-opening the negotiations was Krasin.

The Brest-Litovsk treaty, which imposed yet more burdensome conditions on Russia, was signed on the third of March. Germany, however, did not abide even by the punitive conditions which had been agreed. German troops occupied the Ukraine and advanced into the Crimea and the Caucasus – even into the provinces of Central Russia. Krasin managed to speak to the German Minister of War. The Minister was charming – even sympathetic – but he could only spread his hands in a gesture of powerlessness – the reins of command were now held firmly by General Ludendorf. The Minister advised Krasin to make his way to the front and try to arrange a meeting with the General.

General Ludendorf, that bitter enemy of the Revolution, would fall into a diabolical fury at the mere sight of a red rag. He would not hear of any peace with the Bolsheviks and threatened to remove anyone who attempted to negotiate with Russia or allow the red flag to be raised over the Soviet embassy building in Berlin.

The situation was totally hopeless but, nevertheless, Krasin went to the front line and was granted an audience with the General at which, despite the extreme tension between them, Krasin spoke calmly. He did not speak as a supplicant. He did not try to convince the general that both sides needed peace. From the very beginning of the interview he expressed dissatisfaction with the state of affairs along the demarcation line. He listed the German command's infringements and announced that his country's government would not consider further talks until Germany ceased to violate the treaty. German troops must be withdrawn over the demarcation line and no further assistance must be rendered to (the anti-Bolshevik) General Krasnov. Only when these conditions were met would Soviet Russia consider the development of trade links with Germany.

All the time Krasin was speaking Ludendorf sat upright without moving a muscle. He neither took his eyes off him, nor missed a word he said. When Krasin finished, the General seemed to remain spell-bound for a moment. He complemented Krasin on his "excellent German pronunciation" and on the "style of his exposition", speaking warmly and as if not quite aware of what he was saying. Then he pulled himself together and begun shouting furiously about the madness of the Russians. Finally, now beside himself and choking with rage, he said: "Is it not clear to you Russians that Germany has the power to force you to see reason? All I have to do is to write a few lines right here in front of your eyes – and within a week Russia will be begging us to take more that we demand now!" [22, p. 31] At this point the General saw an ironical smile on the lips of the imposing Russian gentleman before him. Changing his tone abruptly, he added that, if the bankers and industrialists of Germany were interested in trading with Russia, then he was not opposed to such an experiment.

While still engaged in his diplomatic activities, Krasin, in 1918, became People's Commissar for Trade and Industry. He gathered around him the trustworthy elements of the old intelligentsia – scientists, engineers, technologists etc. – that nucleus of expertise from which a whole socialist economy, from heavy industry to the arts, was to grow. Krasin's personal magnetism and audacity "attracted even those who in no way shared the Bolshevik vision, and persuaded them to work for the Revolution." [ibidem, p. 145]

Boris Yefimovich Rumer was content with the way fate had treated him. He worked alongside Krasin and travelled abroad with him on his missions to conclude trade deals between the beleaguered young revolutionary state and those countries which were still trying to maintain an economic blockade.

Boris Yefimovich's work with Krasin brought about his return from Petrograd to Moscow. Home to Moscow! The family was happy! True, the apartment on Kosmodemyansky Pereulok was lost for ever but things were working out so favourably that nobody was too concerned about that.

At first the Rumers stayed with the Briks. Then they moved to the home of the Erenburgs who were spending an extended period abroad. So it came about that in 1918 Yuri was transferred to Moscow University and admitted to the Faculty of Mathematics.

The University of Moscow was still bubbling with mathematical vitality. It would have been easy to imagine that it had been left untouched by the World War, the Revolution, the Civil War and the resulting disruption of normal life. All these things did, of course, have their effect but the work of the Faculty had never ceased for a moment. It numbered some truly remarkable men among its professors: Zhukovsky (TsAGI had already been organised in 1918 on Zhukovsky's initiative!); Chaplygin, Lachtin, Yegorov, among them, too, was the man who has been called "the guiding star of the time, by whose name the whole period in Moscow mathematics is known". The name in question is that of Nikolai Nikolaevich Luzin". Of all the schools which have ever taken the lead in world mathematics, Luzin created one of the most original and prolific – the so-called "Luzitania".

Yuri Borisovich was very much involved in the Luzitania. He considered that it was the school of thinking and of social conduct which shaped his moral outlook as well as his scientific style and aspirations. Its influence lasted all his life – from his first work in mathematics to his decipherment of the genetic code.

This youthful enthusiast was only just 17 years old when he was transported from the devastation and hunger of Petrograd to the equally hungry city of Moscow. He immediately fell in with a high-spirited group of friends. These were people who had everything in common: their youth, their interests, their longing to learn and to pass on their learning. Dressed in felt boots and "déclassé" fur coats for the winter, they would run to see Meyerhold's productions, or to listen for hours to Mayakovsky, standing by the walls in the Polytechnic Museum, or to meet Khlebnikov, to see with their own eyes this "Minister of Global Affairs" carrying all his worldly goods – a pillow case stuffed with verses.

The "Luzitania" had a large crew. If lack of funds made it impossible to create more than one research post, then several young people – recent graduates – would be taken on to share that post. One of Yuri Borisovich's close friends, Lazar Aronovich Lusternik the renowned mathematician, recalled that when he stayed on at the University as a research fellow the post was occupied by ten young people. "The first of these in alphabetical order was Nina Bari, so she was officially appointed. Her salary was then divided into ten parts and shared with the other nine. After the currency reform this amounted to 2 roubles 27 kopecks each." [23, p. 233]

They were not dismayed by the fact that they had to eke out such a meagre living. None of them gave a thought to material blessings. They lived in an exciting world of quest and discovery. They saw things happening which to them with their Tolstoyan and humanistic upbringing seemed harsh, but nevertheless they believed that the new class which had come to power would open the way towards a wonderful new society.

"Life was hard, as a rule, for the young scientists of those early years," said Lazar Aronovich Lusternik, "but youth is always youth. In their case their own physical youth coincided with the youth of the Soviet Republic and of Soviet Science. Wit and humour made life enjoyable. The birth of Soviet Science was attended by noise and fun." [ibidem, p. 214]

All the Luzitanian gatherings, whether the weekly meetings at Luzin's flat where new advances and unsolved questions were discussed, or the "Tatyana's Day" celebrations, or a group excursion out to the former estate of the Trubetskoy family (now the "Uzkoye Sanatorium for Eminent Academics"), were boisterous, jolly occasions. With straight faces people stood up and read comic reports or wickedly satirical papers. Schnirelman wrote fairy-tales in the style of Saltykov-Shchedrin. Many wrote verse. When Lusternik won a state award for his work on topological analysis in 1946, P. L. Kapitza, congratulating him, said that he would certainly get another award for his poetry.

Charades were regular events and jokes about the older generation were always popular. Plays, even operas, were produced. For each special occasion something would be written, usually in a sudden burst of spontaneous creativity. One Easter Yuri Borisovich (now an inveterate thespian, rejoicing in the nickname Lapipid Turandotovich) got together with one of his cronies (Golovachev??) and wrote some parodies. They are perhaps worth quoting here – if only because parody nowadays has deteriorated into mere ridicule. Most parodies contain an epigraph, a line or two taken from the victim's works which either have some unfortunate meaning or no meaning at all out of context. The rest is pure mockery. Real parody hardly needs to contain any direct quotations. It should be immediately obvious who the parodied poet is. Here is The Ballad of the Cottage Cheese as it might have been written by Mayakovsky, Blok, Anna Akhmatova and Vasilii Kamensky:

Vladimir Mayakovsky

It's at you I direct
My withering assonances.
Yes you! Don't protest
Your innocence!
You who live in affluence cossetted,
In bathroom and cosy W.C. closeted,
For now you won't see cottage cheese on sale!
What penalty
For your rumours and tales?
No jails,
No handcuffs,
Just lack of foodstuffs,
For this was the consequence of your jabbering:
State Milk Shop Number Five had more than enough,
a whole roomful.
Until the alerted populace descended.
Now you won't find a spoonful!
Enough of your careless talk, citizens!
Just put the cakes and the jam
On your plate
And try to think straight before blabbering!

Aleksandr Blok

Of famous deeds as well as strangers' staring
All unaware, we wandered in the gloom,
Seeking cottage cheese. We were preparing
To place some on the table in my room.

The hour struck. No curds! Hence your defection
In dead of night just after I retired.
And now I dream of all those sad confections,
Desserts which you so fruitlessly desired.

I dream no more, my love, of you or glory.
The melancholy cake rots on the tray.
My youth is gone – all finished is the story.
With my own hand I throw the thing away.

Anna Akhmatova

Today my body is mournful,
I am feeling tired and worn.
The cottage cheese is running low
And leaving me forlorn.

Quiet as a snail he entered
While rang the Easter bell.
I love the white threads that remind me of
That discarded jacket's lapel;

I beg of you not to reproach me –
There is no cottage cheese, my love.
I shall shrink with despair as on my right hand
I pull my left-hand glove.

Vasilii Kamensky

Cottage cheese, Cottage cheese, Cheese please!
Cottage cheese, cottage cheese, filching cheese
a foe flees.
Cheesed, cheesed, displeased.
Ice, Icicle, bicycle.
Come to the Land of the Cloth of Gold,
Tuck into a pudding made with sour cream.
Who cares?

Osip Brik showed these parodies to Mayakovsky. Mayakovsky read the first and looked puzzled. "What do you mean parodies? Surely this is one of mine?"

Yuri Borisovich was close to Mayakovsky's circle (although, as he always emphasized, he only met Mayakovsky himself four times). Apart from the Briks he had other friends of his own age among these people. One of them was Rita Rait, then a colleague of Mayakovsky's in ROSTA, who was to be a life-long friend.

Rita had intended to devote her life to biology. When she was still a student she was lucky enough to study under Pavlov and, though so young, was allowed to take part in important experiments. This was not enough for her. Commuting between Moscow and the village of Koltushi, "the Home of the Conditioned Reflex", she studied in Bryusov's literary workshop and taught German at the Shanyavsky Institute. She also added translation work to the list.

One day Mayakovsky said to her: "Rita, forget your exams and your dogs and translate "Mysterie-Bouffe" into German. Without dropping any of her other activities, she translated "Mysterie-Bouffe", thought about "The Curative and Preservative Role of Inhibition" and wrote slogans such as: "World Revolution! The latest report! Workers have seized control of factories in Italy! Insurrection in Germany! Uprising in India! Comrades, extend the Revolution ever wider!" This example was on a poster which showed a red flame encircling the globe.

Rita translated "Mysterie-Bouffe" very quickly. The play was staged – and the production was not only the work of the finest directors and designers but also of the young artists of VKhuTeMas who had once had a a widely-reported conversation with Lenin. He asked: "Well then, are you struggling against the Futurists?" They replied: "We are Futurists ourselves!"

This was how these young people lived. They were very poor but they believed that world revolution was not a distant dream but was to be expected the day after tomorrow. Meanwhile they, the children of the new century, were destined to create a new society, a new art, a new science.

They threw themselves into work. They taught in schools and colleges, on workers' evening courses. They taught whatever and whoever had to be taught. They taught soldiers returning from the front and women factory workers. They taught mathematics, German, literature. They organised drama groups. And they still found time to study.

For example, when Lazar Aronovich Lusternik was a student, he taught on the workers' courses: "A significant number of my students were mature people who had some experience of life and knew a greenhorn when they saw one. They behaved condescendingly towards the inexperienced enthusiast who was teaching them. There is a certain amount of exaggeration in the verses which Yuri Borisovich Rumer, then studying at the MGU, wrote about me:

Last night, while teaching workers in my dream, I found
  transfinite numbers were my theme!
It is fair to say, however, that this was the experience which
  led to my popularizing work in mathematics. [23, p.231]

Their life was characterized by this need, not only to study, but also to share their knowledge, to teach, to raise people to a higher level – as well as by their thirst for change and discovery.

Take Pavel Aleksandrov, one of the most brilliant of the Luzitanians, who came close to giving up mathematics altogether. He was destined for a career in music – but at the age of 14 he decided to drop his musical studies and devote himself to mathematics. Of this decision he wrote in his memoirs, that: "it was, as I only realised many years later, probably one of the greatest, if not the greatest mistake I ever made." [24, p. ?] In 1913 he went to Moscow University to study mathematics. He was very soon noticed by Yegorov and Luzin. While still an undergraduate, he wrote his first brilliant paper in which he propounded a new set-theory operation. The sets generated by this operation have become part of the history of mathematics and are known as A-sets (by analogy with Borelli's sets which are called B-sets).

Then, all of a sudden, Pavel gave up mathematics and went off to join his friend Sasha Bogdanov in Novgorod-Seversky. The circle of people who befriended him there were devotees of the theatre. The year was 1918. Pavel Aleksandrov put all thoughts about mathematics out of his mind and plunged into amateur dramatics. In the spring of

1919 they helped to organise the Chernigov Soviet Theatre – Sasha Bogdanov as an actor, Pavel Aleksandrov as a producer – in fact, Pavel took on the formidable task of producing Ibsen's "Ghosts". He also went to Moscow to discuss theatrical developments with Lunacharsky and gave public lectures on Goethe, Ibsen and Dostoyevsky. When the Teacher Training Institute opened in Chernigov Pavel Aleksandrov lectured there on literature.

Denikin's White troops entered Chernigov in the autumn of 1919. They seized Aleksandrov and brought him before a military court where he was found guilty of energetic collaboration with the Bolsheviks and enhancing the popularity of Soviet power. Fortunately the Red forces re-captured Chernigov before the sentence could be carried out and Aleksandrov was freed. He went back to his work in the theatre and to his lecturing. It was at this moment that the punishment of the gods for his betrayal of mathematics fell upon him – he went down with typhus. He was unconscious for two weeks. It was two months before he managed, miraculously, to crawl back from death's door. When he was well again he realized that he had no choice but to return to Moscow and to his mathematics.

"By the late autumn of 1920 I was finally back in Moscow. My reception was just like the Return of the Prodigal Son. V.V. Stepanov and D.Ye. Yegorov welcomed me with particular ardour." [ibidem] Aleksandrov's active and successful mathematical career began all over again. For a long time he considered returning to the theatre but these thoughts remained just thoughts. So it was that the country lost a producer and gained a brilliant mathematician.

Yuri Borisovich recalled the merriment with which the return of Aleksandrov was celebrated. The young Yuri Rumer did not know at that moment that similar adventures would soon befall him, too. He had, while continuing with his undergraduate studies, completed a course in military engineering. He had gained excellent grades and had even been taken on to teach the same course to others. He had also started learning Oriental languages in the General Staff Military Academy. He found these language courses easy; he had already learned a bit of Persian from his brothers. The main attraction was that students on these courses received slightly more generous rations. "I can never forget", Yuri Borisovich used to say , "the taste of frozen potato and wheat porridge. I never could stand the stuff – I even used to refuse it in prison, later on. But in 1921, when everybody was hungry, we language students used to divide a portion into six equal pieces. It was always cold and slimy but, nevertheless, I could always have eaten another piece – but there was no chance of that. Equal shares – that was the sacred law."

While attending these courses Yuri Borisovich soon noticed a man who wore the same Red Army uniform and who had an amazingly large, bulging forehead. No introductions were needed. This man simply walked up to Yuri and asked: "How long did it take you to get the hang of this crazy Persian?" It turned out that despite all his efforts he was making no progress. He had drawn up a strict schedule for himself, laying down how much ground he had to cover in how much time, but he was getting nowhere. The first item on his programme was: "Persian alphabet – one hour."

Yuri could not help laughing when he heard about this hour. No one could possibly master the Persian alphabet in one hour. You cannot, for example, simply learn a letter "a". "A" has to be learned as it appears in various different words, since in combination with different consonants "a" is represented in different ways. If it falls at the end of a word yet another change of form is required. These complications apply to just about the whole alphabet. Besides, as Yuri's elder brothers had always pointed out, you can all too easily lose sight of the wood if you investigate each tree in too much detail: learn something yourself one day, something from someone else the next day; here you will work it out, there you may not. The main thing is to keep on reading, without the dictionary as far as possible, and you will be speaking your new language before you notice it.

That is roughly how Yuri explained his ideas about language learning to his new acquaintance. This man with the high forehead, who became his comrade in that hard and hungry year of 1921, was Sergei Eisenstein. When they were both studying on the language courses Eisenstein was working under Meyerhold, first as an artist, then as a producer in the first workers' theatre organised by "Proletkult". At about the same time he got involved in LEF and Yuri Borisovich often ran into him when visiting the Briks. Not for long however. As the best student on the language courses, Rumer was offered a minor diplomatic post, actually a job as an interpreter, in the embassy which the Soviet state had just opened in Persia. To Yuri Borisovich, who had been nurtured on the poetry of Omar Khayam and Hafiz, this opportunity to spend some time in the Orient was very tempting.

At that time Persia was one of the world's hottest trouble spots. There were constant reports in the newspapers about alarming events such as savage reprisals against the Janghelists – the Persian gangs which brought together elements from the peasantry, the urban poor and the lower-middle class.

There were constant changes of Government as one cruel and reactionary regime succeeded another. Ever since the time, after the October Revolution, when Russian troops had been withdrawn from the occupied Persian territories and the country had been guaranteed full independence, the young land of the Soviets had been proposing co-operation between the two states. But along the way there had been an endless series of obstacles – total confusion in Persia itself, bloodshed, the interference of the Entente. Finally in 1921, thanks to some incredibly hard work, a Soviet-Persian Pact was signed. The Soviet Government revoked all the agreements that had been made between Persia and Tsarist Russia, wrote off all debts, handed over to Persia the Russian discount lending bank and made many other concessions. This did not prevent Reza-Khan, the Persian Minister of War, from arresting the

Soviet ambassador not long afterwards or from committing several other outrages.

The 1921 agreement was signed. The Soviet mission was organised and sent to Persia. Yuri Borisovich, fully convinced that his services there were indispensable, set off for Resht, the capital of the province of Gilyan.

Resht was a terrible experience. The corpses of people who died of hunger or by the knife were left lying for long periods in the filthy, stinking streets. Epidemics of typhus, cholera and trachoma ravaged the country. Life in Resht was an uninterrupted nightmare, a succession of horrific images. Nothing could have been more remote from the poetry of Hafiz. Yuri Borisovich spent more than six months there, living through all kinds of horrors, including the "Shakhsei-Vakhsei" when the Shiites stab themselves in the breast and, streaming with blood, shout "Ali! Ali!" To make matters worse, the Soviet diplomatic staff were not allowed to "pass by on the other side". They were required to give the "Muslim comrades" on-the-spot lectures about the harmful effects of religious observances.

There is no way of knowing how all this would have ended for Yuri Borisovich if he had not caught jaundice. The first attack was severe and the patient's condition continued to be serious. It became clear that there was not much chance of curing him where he was. He had to be sent home. However, because of the shortage of staff, it was decided that the diplomatic bag should go with him to Moscow. So Yuri Borisovich set off on his homeward journey, all alone in a diplomatic carriage with his illness and a pile of secret documents. He was half-conscious when he crossed the border but all was going well until he was held up by the railway police on the outskirts of Saratov. His papers turned out to be not completely in order. Without hesitation the authorities sealed the carriage and arrested Yuri Borisovich. This was, in fact, just what the sick diplomat needed. At least he was admitted to the prison's sick bay and given a certain amount of care there.

Meanwhile, the authorities were sorting things out (not always an easy task at that time): they sent a telegram to Chicherin, the People's Commissar for Foreign Affairs in Moscow, stating that a certain Yu.B. Rumer, who could not produce satisfactory documents (the paper which certified that he was a member of the embassy staff in Persia was not correctly worded) had been apprehended while travelling alone in a diplomatic carriage with important government papers in his possession. The commissars reply was short and to the point. The carriage was to stay sealed until another official arrived; the ailing Rumer was to be despatched to Moscow without delay.

It is a miracle that all this passed off without mishap. In those days he could easily have been shot at any wayside halt – this strange person, thin and yellow with a long nose and feverish eyes, looking suspiciously like a Persian in his weird, brightly-coloured, knee-length trousers, carrying important government papers and travelling with defective documents!

Yuri Borisovich handed over the wretched carriage and hurried to Moscow to make up for the time he had lost while off pursuing romantic dreams and foreign revolutions. Moscow University received back yet another Prodigal Son. This one was emaciated, but wide-eyed with optimism. He arrived at that very time when the star of relativity shone over Luzitania. Pavel Aleksandrov and Pavel Urison had been the first Soviet mathematicians to travel abroad. They had impressed the mathematical schools of Western Europe both with their youthfulness and with their lectures on the new science of topology, which they themselves had created. Now they were back, and lecturing on Einstein's theory of relativity.

Pavel Sergeievich Aleksandrov was to be elected as a corresponding member of the Göttingen Academy of Sciences at the age of 32. Pavel Urison, leaving behind a number of papers which proved to be milestones in the history of world science, was tragically killed in 1924, when he was only 25.

The theory of relativity had a profound effect on the ex-globe-trotter and "diplomat". Now Yuri Borisovich Rumer was sure that he had found his vocation.

Bernhard Riemann

Chapter 6

Where the Crowned Heads of Science Gather

Soon after their first meeting Born announced to Rumer that he had sent his paper to the Göttingen Scientific Society and that Rumer would have to give a talk on it there. The Göttingen Scientific Society was the successor of the old Hanover Academy. At its foundation it had been given a title based on that of the British "Royal Scientific Society". It was composed of actual members and corresponding members and existed, according to Hilbert, with the sole purpose of annoying those who were not members at all. The appearance before the Society was purely a formality. However, before addressing the Academy Rumer would have to pass the real test – his baptism of fire at a meeting of the Mathematics Club.

"This club was a totally informative organisation, having no officials, no permanent members and no funds. Anyone who was interested could simply turn up at a meeting. Nevertheless, such was the general standard of mathematics at Göttingen that the level of discussion was always extremely high." [25 p. 218]

The Club was Hilbert's kingdom – the kingdom, that is, of a man who has left his mark on almost all branches of mathematics. We have Hilbertian space, Hilbert's theorems, Hilbert's transformations, Hilbert's axioms, Hilbertian sets, Hilbert's 23 problems – in short, an enormous mathematical inheritance.

Hilbert not only founded a brilliant school of mathematics. He was also instrumental in the creation of the Göttingen school of physics. As far back as 1905 he and his friend Minkowski had organised a seminar on "The Electrodynamics of Moving Bodies". Strangely enough, this was the very year when Einstein's first famous articles on the special theory of relativity were published – one of them even had the title "Towards an Electrodynamics of Moving Bodies" – but those who took part in the Hilbert-Minkowski seminar were not to hear of them for two more years. In 1908 Minkowski brought the theory of relativity to mathematical completion; he showed that the principle of the constant velocity of light could be stated in purely geometrical terms and created a four-dimensional "world" of space and time which is now called the Minkowskian space. Minkowski's work provided the basis for Einstein's general theory of relativity.

Hilbert was fascinated by the confusing developments which attended the birth of modern physics. Talks given by theoretical physicists became a tradition in the Mathematics Club. One of Hilbert's students wrote: "We mathematicians became uneasy as we listened to the mathematical physicists presenting principle after principle without any supporting evidence, then going on to base all kinds of assertions and consequences on these principles." [ibidem p. 168]

Hilbert's favourite pronouncement, in this connection, was that: "Physics is too complicated for the physicists. The mathematicians had better have a try." [ibidem, p.167] Hilbert, naturally, was perfectly well aware that mathematics alone, despite all its power and potential, would never manage to bring "order into chaos" without a knowledge of the laws of physics or without that special intuition which distinguishes physicists from non-physicists. When he heard about Einstein's successful formulation of the theory of relativity, over which Hermann Minkowski had laboured so mightily, Hilbert said: "Any schoolboy that you could meet on the streets of this capital of mathematics knows more about four-dimensional geometry than Einstein does. Nevertheless, the credit for this achievement belongs to Einstein, not to the mathematicians." [25, p. 186]

Hilbert acquired some physicists as assistants. The first of these was Paul Ewald, who was soon christened "Hilbert's physics teacher". Hilbert had had physicists around him before, Max Born included, but they had concentrated on mathematical problems. It was largely thanks to Hilbert that the 1920s became the Wunderjahre of Göttingen physics. The Mathematics Club, the place "where the crowned heads of science gather", became more and more packed with theoretical physicists. There was no particular distinction between the "crowned heads" and the students, or between the mathematicians and the physicists: if cakes were served then all had a share; if the debate reached no clear conclusion and the need arose to gather in a cafe for "sponge cake and physics", then everybody went along.

The day arrived when Yuri Borisovich Rumer, his name already Germanized to Georg Rumer, gave his talk at the Mathematics Club. Hilbert was not there, but Courant, after listening attentively to Rumer, said at the end: "Can this be a physicist from Russia? I thought their physics was just a matter of glass, metal and wires – a mess, in fact – but this was pure mathematics." Rumer's talk met with approval and was soon published in "The Proceedings of the Göttingen Scientific Society". Elated with his success, Rumer decided that he would show his article to Hilbert.

Not long previously Rumer had had another triumph – he had managed without much trouble to secure the right to use the famous Göttingen library, the Lesezimmer. This library was housed on the third floor of the Auditorienhaus, next to the room where the Mathematics Club held its meetings. After his presentation Rumer went straight to the library.

The Lesezimmer operated an ideal system which had been introduced by Felix Klein. Before his innovations the Göttingen library had, like most other libraries at that time, issued borrowers' tickets. Anyone who wanted one could get a ticket; getting hold of the right book was another matter. To order a book the reader had to supply the exact title, author and catalogue number. Finding a particular book was easy enough if full details were available, but anyone who needed to scan the available literature was up against a hopeless task. Felix Klein changed all this, giving instead free access to all the books which were conveniently arranged under subject headings in alphabetical order. Removal of books from the library was no longer permitted and the reader had to undertake not to replace books after use. They were to be left in a designated spot and two lady librarians did the re-shelving. The library also had a secretary/administrator, Klein having managed to obtain the necessary funding for this post. The Lesezimmer's first secretary was Arnold Sommerfeld, one of Klein's assistants at that time.

Whereas anyone could apply for a borrower's ticket under the old system, it was now necessary to seek a reference from a respectable citizen of the town, who would recommend the applicant and accept responsibility for any loss or damage. When Rumer had first turned up, innocently expecting to use the library, the stern librarians had informed him of this invariable rule. Rumer did not object but went straight to the secretary, a certain Privy Councillor Heims. The librarians had both been brought up in the true German tradition and it therefore occurred to neither of them that this young man might be trying to get round the inconvenient rule.

Rumer, having found his way with such ease into the councillor's presence, greeted him and explained that he wished to use the library but had no respectable citizen to turn to for backing since no one yet knew him apart from Max Born and a few young people. The Privy Councillor looked at the young man with some surprise and started to examine him. Here was a tall, thin fellow with burning black eyes and a smile which never faded; it was almost as if the ends of his mouth were attached to his slightly protruding ears.

"Are you Armenian?" Professor Heims suddenly asked.

"No, Councillor, I am a Jew from Russia."

"That's what I thought. Tell me, do you know anything about Armenian culture? Armenia is part of Russia, after all."

"I wouldn't say I know much. All I know is that the Armenians are Monophysites."

The Councillor's face beamed with a kind of heavenly light:

"Oh, I've been interested in Monophysitism all my life. You have grown up in close contact with the Armenians. Can you tell me something about their religion?"

"I can, Councillor. I can tell you that they believe in the One Nature of Jesus Christ. In Him the Divine Nature is so preponderant that His human nature can be ignored. Christ ("Miadzin" in Armenian) has no human essence." Rumer went on, explaining excitedly, like one inspired, that human reason could not comprehend divine truth, that God is One in Essence and beyond the grasp of disputation. The Armenian Church has its own 'Pope', the Catholicos, whose residence is situated in the same place as Armenia's mother church, the cathedral of Echmiadzin. The name Echmiadzin means 'Descent of the Only-Begotten'. The cathedral was built in the 4th century AD not far from Yerevan at Vagarshapt, the ancient capital, on a site indicated by the Only-Begotten Himself, who appeared in a vision to Gregory the Enlightener. Gregory was later canonized."

Perhaps these were not the young man's exact words but he certainly spoke in this vein and with feeling. Finally the Councillor said: "Herr Rumer, you do have someone who will vouch for you." He called one of the librarians: "This is Doctor Rumer from Moscow. He will be permitted to use the library on my recommendation." Fate had smiled on the young man from Russia. The famous Göttingen library, full of books and journals of which he had before known only the titles, was now open to him.

Now, when the printer's ink hardly dry on his presentation paper, he decided to go straight to David Hilbert's home and see him there. The young man did not suspect that to call in on Hilbert himself without invitation or introduction would be dreadfully tactless. In order to arrange a private visit to David Hilbert one first had to see his assistant Bernais, impress him, explain what the conversation would be about and persuade him to ask Frau Hilbert to allow such a visit.

A maid opened the door. As it happened, Hilbert was at home. Rumer gave the embarrassed maid his visiting card and he was admitted! Hilbert, apparently, said: "Bring him in. I wish to have a look at this Russian gentleman who is not afraid of the Frau Councillor and has got into her home without her knowledge."

Yuri Borisovich often remembered and linked the two portentous encounters: his first meeting with David Hilbert and his first meeting with Albert Einstein. His recollections of his meetings with Einstein have been published in the journal "Nature". An excerpt is reproduced here:

"... The beginning of my time in Göttingen coincided with the beginning of the world economic crisis and the onset of the 'lean years'. Only now, half a century later, now that Born's correspondence with Einstein has been published, have I discovered with what heartfelt concern Born endeavoured to help me, although I was just 'someone from Russia' who had turned up out of the blue. In order to 'beg some money for Rumer', as he put it in one of his letters to Einstein, Born decided to enlist Einstein's good offices and had sent him my paper.

The prophets of the quantum faith had, as a rule, pedagogical tendencies and were usually reputed to be good teachers. The fact that several of these teachers were practically of the same age as their students made their relationship all the more relaxed and lively. The main task for the older generation (in Göttingen this meant Born) was to select likely recruits – who would then be passed on to the young teachers. Einstein himself was never regarded as a great teacher. His enormous inner concentration was an almost insuperable barrier facing any person who tried to gain an insight into his ideas .... As his letters to Born testify, Einstein repeatedly expressed his wish to 'find manpower' to help with calculations.

Born was surrounded by a crowd of young and creative people and was always on the look-out for suitable assistants for Einstein. He thought he might 'try me for size' and sent a copy of my paper to Einstein along with a covering letter in which he recommended me as an 'ideal assistant'.

There was not much chance that anything would come of this letter, as Einstein did not usually read other people's work. However, Einstein's friend Professor Pavel Sigismundovich Ehrenfest, whose links with Russia went back many years, was deeply concerned about the fate of all young Russian physicists who ventured abroad. Ehrenfest took up my case. No doubt Born played a vital part, too. In any case, in the December I received a telegram from Berlin. It was from Ehrenfest: 'Come. Einstein will see you.' On the heels of the telegram came a postal order for 200 guilders for my fare – which was very welcome." [18, p. 109]

From the moment he received the telegram from Ehrenfest Rumer was in a state of agitation. He thought constantly about the coming meeting, carefully considering what to say but the words would not fall into the right order.

"At the beginning of December 1929 I arrived in Berlin and went straight to see Einstein. I did not have to wait long. The door of the sitting room opened and in he came. He approached and offered his hand introducing himself simply as 'Einstein'. I answered: 'Good morning, Herr Professor' and these everyday words immediately settled my nerves, making the great man seem more like one of those Göttingen professors with whom I was already at ease. Einstein's hands stick in my memory: they were more like the hands of a brick-layer than the hands of an academic. I was soon to see such hands again when I met two other men – Landau, who was then still a junior, and Dirac, who had already achieved greatness.

Then Ehrenfest came in. I had never met him, either. He greeted me in that inimitable 'Ehrenfest-Russisch' which, so it seemed to me, gave him such pleasure. We went off to an attic room with a low wooden ceiling – Einstein's study . After an hour and a half of conversation, for which my work served only as the starting point, I was exhausted. I remember being very surprised that my companions were both still fresh and responsive, apparently not tired at all. Ehrenfest and I went back down to the sitting room. Einstein's wife came in and very kindly invited us to stay for dinner. I accepted, but Ehrenfest said: 'No, don't stay. I shall have to talk to Einstein about you over dinner and you might be in the way.' Then Ehrenfest went off to talk to Einstein, presumably about my future." [18, p. 110]

While the meeting with the greatest of all physicists was to decide the young man's fate and so could hardly fail to unnerve him, the meeting with David Hilbert, which Rumer arranged for himself, began in a spirit of boyish mischief. He went merrily and proudly to see Hilbert, with the curiosity of a kitten which is not frightened of any other beast on earth. However, when he heard Hilbert's voice through the half-open door, saying: "... I wish to have a look at this Russian gentleman ...", he was seized by panic. The realization hit him: "Hilbert! My God, what have I done? What shall I do now?" He would now enter that room and Hilbert would ask: "Well, what brings you to me, young man?" How would he answer that? That he wanted to show Hilbert his paper on the generalisation of the general theory of relativity on the basis of the Gauss-Codazzi equations? God, what nonsense! He did not remember walking into Hilbert's study. He seemed somehow to materialize in front of the desk at which sat a man with an absolutely white beard and a large, handsome forehead. The eyes were the most striking feature of this face; they were huge, blue and clear. They surveyed the intruder with a touch of irony but with obvious interest.

"What has brought you to Göttingen, Herr ...?"

"Rumer", Yuri Borisovich heard himself say.

"Herr Rumer?"

"Extra Göttingen non est vita". ("There is no life outside Göttingen.") Yuri Borisovich found himself recalling the Latin motto which adorned the wall of the Ratskeller and again he thought: "What rubbish!" But aloud he said: "The revolution in physics".

"Well, anyone can see straight away that you are from Russia. You Russians are all great experts on revolution. Not a bad thing. Lobachevsky was a revolutionary, too."

"Yes, Herr Councillor, but Gauss couldn't be called a revolutionary." (God, what was he talking about? Why bring Gauss into it?)

"How can you say that? Can you imagine where we would be without Gauss? Without all that he did?"

"One cannot imagine it, Herr Councillor, but he was not a revolutionary. He never overturned anything – he was peacefully creative – he never upset the established order."

Hilbert became animated. He started to talk about German mathematics, about how it had long remained within the confines of the trivium and how it only started to blossom when Gauss came on the scene, how Gauss placed the German school in mathematics on an equal footing with the Anglo-French school. No – Gauss was the king of mathematics.

"Indeed, Herr Councillor, I would say that this title is more befitting than any other. He is precisely what you say – a king, not a revolutionary. Riemann, however – now he was a revolutionary." ("Damn this revolution – I'll be thrown out for it." thought Rumer).

Hilbert, however, suddenly broke out into ringing, child-like laughter. This unexpected mirth surprised and annoyed Rumer:

"Herr Professor, Gauss made huge advances and his fame is thereby assured. But the fact that he once betrayed both the truth and the people who needed his support will always be held against him."

This started a heated dispute between the young man who had "walked in off the street" and the great David Hilbert concerning truth in science. This led to a long conversation about the amazing story of the birth of non-Euclidian geometry – a story which unfolded within the ancient walls of Göttingen and of distant Kazan. Then they talked about the unique genius of Lobachevsky, about the tragic end of the young Janos Boljai and about Riemann.

Non-Euclidian geometry was in fact the truth which the King of Mathematics betrayed. Was it really treachery? History cannot be so emphatically certain as our young hero was. History is fact. For a century there have been bitter arguments about the facts of this case. One can juxtapose them, weigh them, come to logical conclusions and make judgements but it is hardly possible to arrive at the actual root causes of the situation which developed.

From his youth on, and all through his long life, all through the years of his intellectual achievement, the King of Mathematics was always haunted by a problem arising from Euclid's fifth postulate. Euclid created the geometry that we know today, the geometry that we learn in school, on the basis of five postulates which he himself formulated. The first of these, for example, states that it is possible to draw a straight line from any point to any other point. The next three are equally simple and obvious and do not give rise to any doubts. The fifth postulate, however, is more like a theorem. It leads to the assertion that two parallel lines can never cross and consequently, that the sum of the internal angles of a triangle is always 180°. Euclid himself obviously first formulated the fifth postulate in the form of a theorem. After painstaking attempts to prove it, Euclid was obliged to call this theorem the fifth postulate.

No one had ever managed to prove Euclid's fifth postulate. What, then, if it were abandoned altogether? What if it were conceivable that the angles of a triangle amounted to less or more than 180°?

Let us imagine a sphere or, better, a globe. Let us draw a triangle on it. One of its vertices is on the pole, one on the San Tomé Islands in the Gulf of Guinea (on the 0° meridian) and the third on the Galapagos Islands in the Pacific Ocean – i.e. at a longitude of 90° east of Greenwich. We thus construct a triangle the sides of which meet at angles of 90° and the sum of the internal angles of the triangle is 270°! A sphere has positive curvature. If we take a surface with negative curvature – which is difficult to visualize, but it may help to think of an old-fashioned hour-glass – then the angles of a triangle on such a surface add up to less than 180°. Moreover, the sum of the angles will decrease as the curvature of the surface increases. And what if our whole universe is curved? In that case Euclid's fifth postulate is false or, at least, is valid only for flat space. There is a need for a geometry which is free of this limitation and so valid for curved space.

Now that such a geometry exists, now that we have the theory of relativity and we know how closely connected are the curvature of space and the force of gravity, all this seems obvious. But then, a century and a half ago, it was by no means so simple. The first person to give serious attention to this problem, and even then not with the full force of his intellect, was Karl Friedrich Gauss. These matters concerned him all his life and gave him no peace. He came within an ace of creating the new geometry but lacked the conviction to publish the results of the work in which he had been so deeply engrossed. The world only found out about Gauss's work in this field five years after his death when their vow of silence was lifted from his friends and they started to publish his letters.

"The assumption that the sum of the three angles of a triangle can be less than 180° leads to a geometry which is totally different from that of Euclid. This geometry is perfectly consistent and I have developed it absolutely to my own satisfaction. I am able to solve any problem in this geometry except that I cannot define a certain constant the value of which a priori cannot be established. The propositions of this geometry may seem somewhat paradoxical and to a person not familiar with it, even absurd, but calm and rigourous reflection reveals that they contain nothing which is impossible.... However, in any case, you must regard this as a private communication which must definitely not be published." [26, p. 187]

Gauss's letter, from which this short excerpt is taken, is addressed to Taurinus and dated 1824. What it contains is, in fact, a quick sketch of non-Euclidian geometry. In 1829 Gauss wrote to Bessel on the same subject: "It will probably take me a long time to write up my lengthy investigations into this matter in a form which can be published. It is even possible that I shall not pluck up the courage to do this in my lifetime because I am afraid of the howls that the Beotists will raise if I state my views in full." [26, p. 200]

Gauss – afraid? Karl Friedrich Gauss, the King of Mathematics, whose word was law, afraid to speak out about a scientific truth which he himself had long acknowledged? Though it seems barely believable, the fact remains: Gauss not only refused to publish the results of his investigations, he also swore his friends to strict secrecy concerning his new geometry.

Gauss chanced to become involved in a geodesic survey – or perhaps it was not entirely by chance that he was given the task of mapping the State of Hanover. Was this, after all, a fit occupation for the King of Mathematics? There is no doubt that Gauss could easily have refused this commission if he had not had his own reasons for accepting. There were indeed reasons – which he kept secret! Gauss designed a precise measuring instrument for the survey. The arrangement was that specially detached army officers would carry out the survey work, after which Gauss would perform the necessary calculations. The result was an extremely accurate map of Hanover. While this work was going on Gauss was working on his own secret project. The most closely guarded secret was that Gauss himself measured the angles during the triangulation work. He was hoping to find that the sums of the angles of the huge triangles involved would not be the expected 180°. Unfortunately for Gauss, the State of Hanover, great though it was, did not extend far enough to demonstrate the operation of the laws of non-Euclidian geometry. It was, for all practical purposes, flat.

At about the time when Gauss was measuring the angles of his triangles laid out on the Land of Hanover, in far-off Kazan Nikolai Lobachevsky was lecturing on "imaginary" geometry. The lecture was delivered on the 12th of February, 1826 at a meeting of the university council. Lobachevsky did not finish his lecture. There was no point in trying to, as no-one present could understand it. Lobachevsky was derided. Such was the beginning and so it continued until the end. For 30 years Lobachevsky developed his geometry and right up to his last work "Pangeometry", which he dictated to his students when he was already blind, he was rewarded only with ridicule and humiliation, public attacks and anonymous letters. But he turned out to be braver than Privy Councillor Gauss. He published all his works, beginning with the lecture which had been such a disaster, and proved for all the ages to come that a scholar needs not only talent, but also courage.

Lobachevsky's first pamphlet "On the Principles of Geometry" appeared in the "Kazan Herald" in 1829. A German translation of this work came to the attention of Gauss in Berlin 11 years later. Gauss's letters show how deeply affected he was by the paper. He immediately responded to it – not, however, with a public acknowledgement of the new geometry, but with something quite different. He successfully nominated Lobachevsky for acceptance as a corresponding member of the Göttingen Academy of Sciences. In his nomination speech Gauss, just as would be normal nowadays, extolled Lobachevsky's great services to mathematics. About non-Euclidean geometry, however, he did not say a single word. It was as if it did not exist. Neither did Gauss ever write a word to Lobachevsky himself, at that time or later. Lobachevsky waited, perplexed. After all, one word from Gauss would have been enough to save him from the endless taunts and indignities and to win recognition for his life's work. Lobachevsky's pride would not permit him to make the first move in any correspondence with Gauss. Gauss, for his part, remained silent, although throughout those 30 years (Lobachevsky survived Gauss by only one year) he kept track of Lobachevsky's work.

In his extreme old age Gauss started studying Russian in order, as he wrote to the astronomer Enke, "... to read more of the works of this brilliant mathematician." [ibidem, p. 232]

Although Lobachevsky continually published his work both in German and in French journals, Gauss justifies his study of Russian very simply in that same letter to Enke: "The publications of Kazan University contain a large number of his works." But for 30 years he made no attempt to contact Lobachevsky, even by letter. To his friends he wrote: "Lobachevsky calls it (the new geometry) imaginary geometry. You are aware that I have held the same convictions for 34 years (since 1792) and that, therefore, as far as basic principles are concerned I have found nothing new in the work of Lobachevsky. However, he follows a path of development which is different from the one I took myself and he follows it with great skill, in the true spirit of geometry." [26, p. 235] He "held the same convictions" but never said so publicly and never sent a single word of approval or encouragement to Lobachevsky himself!

Meanwhile, Lobachevsky lived and worked. In 1816, at the age of 24, Lobachevsky was appointed "professor extraordinary". Several times he was elected as dean and in 1827 he became rector of Kazan University. When he took on the post the university was "in a state of total decay as regards both learning and morale" and presented "a pitiful and shameful spectacle". Soon after his appointment as rector Lobachevsky wrote: "Just as I previously could only suppose, now I can say from personal experience that the responsibilities of the rector are enormous... I am sure that you will not take my words to be an attempt to aggrandise my labours in your eyes. Neither do I wish to have too little confidence in myself. After all, my disposition and personal rule are such as to prevent me falling into despondency and regret over things that I can do nothing to help. I consider diffidence to be pardonable before any decision is yet made but when the mind is made up then there is no more excuse for low spirits.

No doubt you noticed how long I hesitated and even sought to decline the post; now I must be firm and strive with all my strength." [28, p. 75]

Such a man he was in everything – one who "strove with his strength". He lectured on almost all branches of mathematics, experimental physics, astronomy, hydraulics and hydrostatics. He built a chemistry laboratory in the university as well as a physics laboratory, an astronomical observatory and an anatomical dissecting theatre complete with clinics – all this while making a conscientious study of building and architecture. He collected a brilliant library and built a vaulted hall to house it. He was one of the founders of the Kazan Economics Society.

Lobachevsky was a true rationaliser not only in scholarly and social matters but also in the running of his own household. He laid out a garden on his own estate where he used his own grass-sowing system, bred pedigree cattle and even won a silver medal for his Merino wool, which he entered at the St Petersburg agricultural exhibition. He built a dam and a water mill and invented a new type of bee-hive. Admittedly, in these projects he met with setbacks which led finally to ruin, but to the end of his days he was true to himself: "... my disposition and personal rule are such as to prevent me falling into despondency and regret over things that I can do nothing to help".

So it was with his geometry. He worked on it ceaselessly in complete academic isolation and brought it to such a state of completion that his successor had only to apply it. Only one person knew that Lobachevsky was not alone – and that person was Gauss. Gauss knew this not only because he himself had arrived at a non-Euclidean geometry and fully accepted all Lobachevsky's works, but also because Gauss had another work in his possession – a complete and consistent exposition of the foundations of the new geometry by a young Hungarian mathematician called Janos Boljai. Boljai stands alongside Lobachevsky in the history of science as a creator of non-Euclidean geometry. But this is thanks to his successors. Janos was not yet 20 when he created his geometry. His father, Farhas Boljai, was a close friend of Gauss and when they were young had worked with him on parallel lines. Now with pleas, now with dire warnings, the father tried to steer the son away from this "impenetrable darkness". But Janos conquered the "darkness". He wrote to his father: "It is true that I have not yet reached the goal but I have obtained some remarkable results – I have created a whole new world from nothing!" [29, p. 18]

It took Janos a long time to write his paper. The father did not accept the son's work but agreed to insert it as a supplement in his own mathematical textbook "Tentamen". He sent his "Tentamen" to Gauss with a request that the latter should read and evaluate Janos's "Appendix". For a long time there was no reply from Gauss. Then it came. It turned out to be a tragic blow for Janos. It transpired that Janos had produced nothing new, that Gauss had obtained the same results 30 years earlier. But priority was not the issue. It did not matter who was first. Janos never imagined that he was a match for Gauss. What mattered was that truth, scientific truth, should be acknowledged – and upon whom was it more incumbent to acknowledge the truth than upon the world's leading mathematician? Janos Boljai hoped and waited for the day when he would receive Gauss's approval and recognition. He waited until the years of his senility, but in vain. He fell victim to a serious mental illness and died, lonely and unknown. He was buried in a common, unmarked grave, his only epitaph an entry in the church register: "His life was wasted". [30, p. 92]

In his later years Gauss again had to go through intense and deep suffering as a result of his secret. This time the other person involved was Bernhard Riemann.

There have been many great scholars – more of them than there are peaks on the Earth's many mountain ranges. Every peak is magnificent and lovely and it is impossible to imagine how the landscape which it dominates would look without it. However, if we so wish we can measure these peaks in metres and compare one with another. The fact that some will prove to be not as high as others does not lessen their significance. Nevertheless, among them Everest will stand out. Try, on the other hand, to find, even for yourself, the highest peak among scientists – it is impossible. But if there were some kind of unimaginable yardstick by which scholars could be measured as mountains are measured in metres then, possibly, Riemann would be marked as the summit.

Bernhard Riemann was born into a large, poor family in 1826. At the age of 25 he defended his doctor's dissertation at Göttingen University. It was on the theory of the functions of the complex variable. Gauss was delighted with this piece of work. But Riemann did not get the post in the observatory (where Gauss was director) for which he hoped. Riemann's material circumstances were lamentable. He was half-starved. Left without means of support after obtaining his doctorate, he applied for the post of "privat-dozent". The selection procedure required submission of three pieces of work, one of which would be chosen as a "trial" lecture. Riemann worked for a year on this and as required submitted three topics to the collegium. Two of them dealt with problems of contemporary mathematics, the third with the foundations of geometry. Karl Friedrich Gauss was already 75 years old. For so many years he had held onto the secret of non-Euclidean geometry, keeping it locked away in an impregnable fortress. For the trial lecture, Gauss chose the third of Riemann's themes. Gauss's close friend Wilhelm Weber, who had worked with him to create the first wireless telegraph, put it like this: "... he (Gauss) desperately wanted to hear how such a young man would manage to emerge from such a difficult contest."

Riemann exceeded all Gauss's expectations. It became clear that, while Lobachevsky and Boljai had constructed a new geometry by re-jecting Euclid's fifth postulate, this young man had freed himself completely from the very bases of Euclidean geometry and had created a new geometry founded entirely on his own principles. It turned out that there were whole classes of non-Euclidean geometries.

Riemann's trial lecture left the audience stunned. A spirit of forbidding conservatism and formal rigidity held sway in mathematics at that time. Not a single piece of work, not a single assertion was accepted by the Mathematical Society without solid proof. Proof was the Law. This being the case, Riemann's first address, his trial lecture, was a real challenge. His exposition was such as to require neither blackboard nor chalk. Basically he explained his ideas for a new geometry, putting forward his own principles without thinking of proof.

Those sitting in the hall were shaken simply by the form of the exposition, to say nothing of its content, which nobody understood, except for one man. It was at that man alone that the lecture was aimed. Karl Friedrich Gauss understood both its meaning and its significance. He sat, moreover, in a state of "utter amazement" about which only his close friends found out. Riemann had no inkling of Gauss's consternation. He only saw how Gauss got up in silence at the end of the lecture and walked towards the door.

However, neither then nor thereafter at any time in the course of his short life was Riemann concerned about the reaction of an audience or about public opinion in general, not even the opinion of the world's greatest mathematician. He made discovery after discovery, gave his free-thinking lectures and his imagination knew no limit. Everything that Riemann touched acquired a deep meaning and the Law of Proof retreated before him. "Riemann's zeta-function", which is the foundation stone of analytical number theory, is well known to all mathematicians and physicists. Riemann introduced and described it practically without proof. The assumption regarding the location of the "non-trivial" zeros of the zeta-function remains to this day an unproven hypothesis. It was to be number nine in Hilbert's list of 23 problems. It is well known that Hilbert, when asked for his opinion as to what was the most important problem in mathematics, replied: "The problem of the zeros of the zeta-function and not only in mathematics. It is the most important problem in the world." [25, p. 124]

There is a legend which goes like this. Some mathematicians came in an aggressive frame of mind to see Riemann. They demanded an answer to their question. Where are the zeros of the zeta-function? Perhaps they are not there at all? "They are definitely there", Riemann is supposed to have said, "but how should I know where they are? When I die I shall, of course, go to Heaven. I shall go up to the Lord God and ask Him 'Lord, where are the zeros of the zeta-function?' And the Lord God will reply: 'I haven't got a clue'."

History has honoured Riemann's achievements but there is another aspect of Riemann's work which long remained obscure. Pure mathematician invariably becomes fascinated by other matters. In Riemann's case it was something like a unified field theory. He was interested in the connection between light and gravity and saw the forces of nature as being the main causes for the curvature of the space we live in. Because Riemann died young we are left with Riemann the mathematician – while his research in "synthesising physics" has been neglected. Naturally, these sketches of unified field theories may amuse the reader and have a purely historical, rather than any scientific value. Maxwell's electromagnetic theory was still to come. Even that would not have been sufficient. Decades had to pass, Albert Einstein had to be born, before the link between the new geometrical ideas and the physical world could be confirmed.

Riemann's genius looks all the greater for that. This is what he wrote in a letter to his brother: "I have once again taken up my research into the links between electricity, galvanism, light and gravity and have made such progress that I shall definitely be able to publish the latest version of my work. By the way, I now know for certain that Gauss has been working on the very same questions for many years and has now told several of his friends including Weber, swearing them to silence as always. I hope it is not too late to establish the fact that I made these discoveries quite independently of Gauss. I write to you without any fear that you might reproach me for my inappropriate arrogance." [31, p. 44]

Five years later he wrote to his sister: "I have presented my discovery of the connection between electricity and light to the Scientific Society here (i.e. in Göttingen). Judging by rumours that have reached me, one has to conclude that Gauss has constructed a theory of this connection which is different from mine and has told his close acquaintances about it. I am firmly convinced, however, that my theory is the correct one and will be recognized as such before many years have passed." [ibidem]

Gauss never gave a thought to the link between the geometry of space and electricity, galvanism, light and gravity. This is a 20th century problem. It is still unsolved. It is precisely this question which is now called the "front line" of fundamental science when the forward area of the massive front of modern research has to be defined.

"Mr Rumer, let us not pass judgement on the great men of this world. We must take them as they were. Come to our soirees. You will see Noether solving logical problems which everybody (myself included) thinks up especially for her!!"

And so Rumer went home. Three months later Hilbert would reach his 68th birthday. This was the official retirement age. The general opinion was that his only possible successor was Hermann Weyl.

But would Weyl agree to move to Göttingen? Nobody knew. Once, ten years before, Weyl had refused an invitation to come to Göttingen. This time, too, Weyl hesitated but finally he made up his mind and in the spring of 1930 the town of Göttingen welcomed Weyl as Hilbert's successor. Our young man would not, of course, have believed a word of it if he had been told at that moment that in less than two years the article "A Basis for Independent Invariants in Vector Space" written by three authors – Hermann Weyl, Rumer and Edward Teller – would appear in the "Göttinger Nachrichten". He would simply not have believed that he would be a co-author along-side the occupant of Hilbert's chair, which had also once been the chair held by Karl Friedrich Gauss. As for Edward Teller, future fame awaited that modest and then still unknown young man – the dubious fame of being the father of the American hydrogen bomb.

Rumer and Max Born

Chapter 7

Professor Born's Circle of Friends

"Everything was falling into place quite smoothly", wrote Yuri Borisovich, "All that I now had to do was to find some means of subsistence. It soon became clear that this was far from being a simple matter.

Nowadays it is difficult to imagine the difficulties facing even a man like Born when he tried to arrange a post for someone whose services he needed. At that time physicists were either paid for teaching physics or received scholarships in order to study it. It was not accepted practice to pay salaries to research assistants. Moreover, my arrival in Göttingen and the first few months of my stay there coincided with the onset of the world economic crisis and the beginning of the "lean years". [18, p. 108]

The first financial assistance that Rumer received came quite accidentally. Some money arrived from a banking institution known all over Europe – that of Baron Warburg.

Yuri Rumer and his young wife Mila, on arrival from Moscow, had first gone to Oldenburg. There were two reasons for this: firstly, Boris Yefimovich Rumer, approving his son's decision to go to Germany and still hoping that he would turn out to be a good engineer, had persuaded him to go to the Higher Polytechnical School at Oldenburg; secondly, there was an excellent school of rhythmic gymnastics at Oldenburg where Milochka hoped to study. Such schools were being set up all over Europe at that time. Rhythmic gymnastics had been invented by the Swiss musician Jacques

Dalkroz. The famous Dalkroz system, in which movement was linked harmoniously with music, was very popular in aristocratic circles.

Milochka Rumer had been an enthusiastic devotee of this system even in Riga and here, in Oldenburg, she managed to get into the school run by Berning, a pupil and friend of Dalkroz himself. She made friends with Berning's daughters Gerta and Matilda. Actually, of the two sisters only Gerta was really a close friend. She was the more intelligent, the more exuberant and the kinder of the two. However, Gerta by then only appeared in Oldenburg when visiting her family. She was studying medicine at the University of Göttingen. When Rumer arrived in Göttingen Gerta Berning was the only person he knew in the whole town.

Gerta had a close friend called Renata Menkeberg who also sometimes attended Gerta's father's rhythmic gymnastics classes. The Menkebergs were a rich , high-society family. Renata's grandfather was the Burgomaster of Hamburg and one of the main streets of that city bears his name to this day. From time to time Renata Menkeberg visited her friend Gerta in Göttingen. She would arrive like a whirlwind and life would be jolted out of its usual rut: there would be parties, walks in the dark, potatoes baked in pottery kilns in the studios of artists.

In the course of one of these visits Rumer, too, was caught up in the merry whirlpool around Renata Menkeberg. Gerta presented Georg Rumer as a future celebrity, for the moment lacking the means to maintain himself. It could not even have occurred to this pretty, vivacious girl, who had been brought up in luxury, that she might meet a person who had no money. "We'll soon put that right", Renata said gaily, "Three days from now I am going back to Hamburg and then we're off to the mountains. Baron Warburg has arranged a little skiing trip. I'll talk to the old man. What's a thousand to him?" Then she went on just as gaily to babble about some kind of kimono which was coming into fashion. Rumer did not give a second thought to this "thousand", so superficial did all this chatter seem to him.

Three weeks later Renata Menkeberg turned up again in Göttingen. She gave a light-hearted, dramatised account of her negotiations with Warburg, playing in turn the parts of the Baron and of herself. She was a magnificent mimic. "Leaning" lightly on a non-existent ski pole, she gasped and deepened her voice when she portrayed the out-of-breath old Baron.

"I say, Baron, you couldn't possibly let me have a little money, could you?"

"What, here in the mountains?"

"No, Baron. I know of a very gifted ... doctor.

At the words "very gifted" Renata became the Baron and gave a shudder.

"I am embarrassed even to ask you for such a sum, but I have to".

"How much will your doctor cost?"

"A thousand Marks would probably do for a start." Renata gave a sigh of relief for the Baron and mumbled:

"We'll see to that, my dear".

The money that Rumer received from Baron Warburg astounded Born no less than it astounded Rumer himself. A penniless foreigner from some weird country had turned out to be so outstanding that someone had tackled Baron Warburg on his behalf – and successfully, too. Born even mentioned this in one of his letters to Einstein (the letter is dated 13/11/1920): "... Herr Rumer is here in Göttingen. Baron Warburg in Hamburg has sent him some money to enable him to carry on with his work for a while." [2, p. 12]

It would appear that Born deliberately slipped this insignificant remark about Rumer into a letter which contained no other reference to him. He had already written to Einstein in August about "a young man from Russia". He had sent a copy of Rumer's thesis with a request that Einstein should evaluate it and do what he could to help.

Einstein did not send a reply to Born on this occasion but wrote directly to Rumer, saying that he did not like his thesis and that he did not find this topic at all interesting, anyway. Rumer was, of course, cast down by Einstein's disappointing response, but not crushed.

Then Born wrote to Ehrenfest. This is how Pavel Sigismundovich Ehrenfest entered Rumer's life. Born sent Rumer's thesis to Ehrenfest, along with a copy of his own letter of 12 August to Einstein, and asked for his assessment. At his very next meeting with Einstein, Ehrenfest asked him casually what he thought about Rumer's work. "Who is that?" asked Einstein, "I don't know his work. Tell me what you find interesting about it." Ehrenfest told him.

"Why didn't they send this work to me?"

"What do you mean? Born sent it – complete with a long covering letter. Look, I've got a copy of it. And you replied that you were not interested".

"Oh, don't you and Born realise that I haven't got time to read all the papers that people send me? Tell Born to send this young man to see me."

Ehrenfest played a very special part in the development of modern physics. He had a great critical mind. Einstein wrote about him: "His greatness lay in his extraordinarily well-developed ability to grasp the very essence of a theoretical concept and so to strip a theory of its mathematical trappings that the simple underlying idea was revealed with complete clarity. It was thanks to this ability that he was without equal as a teacher." [32, p. 131]

Whenever any new paper was published everybody would ask what Ehrenfest had said about it. It was considered that if Ehrenfest had given it his approval then one had to read it; if not – then it was not worth reading. It was Ehrenfest's fate to become a close friend of Born and Einstein, Bohr and Pauli, Planck, Dirac ... . In his recollections of Ehrenfest Joffe wrote: "An Ehrenfest seminar would attract scholars from all over the place. To deliver a paper at one of his seminars and answer the questions that followed was a great honour, and the subject matter of the paper itself would be enriched as dozens of hitherto unrealized possibilities were opened up." [33, p. 43]

In one of his letters Einstein wrote: "Pavel! I value your friendship so highly! I need your friendship far more than you need mine." [33, p. 44]

Immediately after his conversation with Einstein about Rumer, Ehrenfest sent the young man a telegram saying that Einstein was expecting him, and a little money just in case.

On 14 December Einstein wrote to Born: "... I liked Mr. Rumer very much. His idea about the attraction of multidimensional sets is original and well formulated. The weakness lies in the fact that the laws thus revealed are not complete and the path leading towards their logical substantiation and completeness cannot yet be discerned.

In any case, it would be a good thing if this young man could be given the opportunity to go further in his scientific work. The best thing, of course, would be a post that left him enough free time for his own studies ... . Would it not in such cases be possible to find teaching posts in grammar schools or other posts which offer a modest workload for a modest salary? This is definitely a better solution than scholarships granted for a fixed period. The problem is that the stork which brings the child of the mind is a very wild bird which pays no heed to delivery dates." [2, p. 15]

Born answered Einstein's letter without delay (his reply is dated 19 December):

"Dear Einstein, I am very glad to hear that you wish to accept Herr Rumer. The idea of finding him a post which would leave him enough time for his scientific work is, of course, very sound in theory – but presents practical difficulties. The ideal solution would be a part-time teaching post in a grammar school but, of course, that is very difficult to arrange. He would probably have to complete several years' training first. My own contacts in the ministry are not such as would enable me to ask any favours in this connection. Perhaps I shall have to ask you to use your influence. If the weight of your name could be brought to bear, this practical problem might be solved, much to the benefit of our young people. Could you not ask to see Richter, the head of the ministry, and explain the whole thing to him?

However, these are all dreams of future possibilities – no use at all to our Mr. Rumer. By the way, Rumer has some practical qualifications. He completed a course at the technical college in Oldenburg. He could do some kind of practical work but with unemployment as it is in Germany at the moment a foreigner has absolutely no chance of getting a job.

For the moment, I think, there is no other way out but to provide him with a grant, at least for one year. My wife tells me that you and Ehrenfest were going to ask the Rockefeller Foundation for help. Please, could I ask you to write to Mr. Tisdal (Rockefeller Foundation, Rue de la Bom, Paris) and ask for a grant to enable Rumer to work for a year with you, or perhaps with me or wherever seems best. Say that Ehrenfest and I support this application." [2, p. 15]

In his commentary on the letters just quoted Born wrote: "Einstein always said that the search for knowledge should not be the work that the student depends on for survival. Research should be an independent pursuit. He himself wrote his first major article when he was earning his living as a clerk in the Swiss Patents Office in Bern. He thought that was the only way to maintain one's intellectual independence. His suggestion that Rumer should try to find a part-time teaching post was in accord with this principle.

He failed to take into account the organisational inertia that affects almost every profession and the importance which practically every person attaches to the work he or she does – and without which it would be impossible to work up the necessary professional enthusiasm. In order to succeed in science when it is only a leisuretime activity one has to be Einstein." [2, p. 17]

Rumer had no idea at that time to what lengths Born was going in the attempt to secure him an income. "Only now, half a century later," wrote Yuri Borisovich, "now that Born's correspondence with Einstein has been published, have I found out about Born's whole-hearted caring commitment to me, 'the man from Russia' who had turned up completely unannounced." [18, p. 111]

Born wrote and begged and managed to arrange for Rumer a meeting with Einstein. This meeting, as we know, passed off success-fully. Rumer, however, did not know this at the time and as he was leaving Berlin he asked Ehrenfest how things stood. Ehrenfest replied: For the moment you are to carry on working in Göttingen with Born."

"I realise now, of course," said Yuri Borisovich, "that the purpose of the meeting with Einstein which Born arranged for me was to assess my "psychological compatibility with the great man."

The question about my compatibility turned out to have been answered in the affirmative, as can be judged from Einstein's letter to Born on 14 December 1929 but at this very moment when my fate had already been "counted out and weighted", I, suspecting nothing of all this, had returned to Göttingen, where I had teachers such as Heitler and Weisskopf, and plunged back into the bubbling pot in that "Göttingen quantum kitchen" which fascinated me more and more." [18, p. 111]

The barefoot shepherdess smiled serenely, confidently awaiting further "quantum kisses" – and there was not a young man who did not dream of kissing her.

Practically all of Born's young people lived in Frau Grönau's pension – the so-called "quantum boarding house". This establish-ment was made famous by its then far from famous lodgers. Heitler lived there, as did Nordheim, Chandrasekar, Weisskopf, Eduard Teller and Max Delbrück (who later deserted physics and received a Nobel prize for his work in genetics). There were Japanese, Indians and Chinese there, too. The Chinese all lived in one room, wore identical clothes and regularly got up at 5 a.m. to sweep the streets of Göttingen, without any offer of money for their work.

This was the lodging house to which Rumer moved when Mila came to him from Oldenburg. Mila was a pianist so they hired a grand piano and took two rooms. Gatherings now took place more often than not in the apartment where "the Rumers and their grand piano" resided.

The people in the lodging house were a single household, bound together by genuine friendship and common interests. They all went to the cinema together and were even able to persuade the cinema owner to show them the "Threepenny Opera" twice over at one sitting. For them this was an event comparable to tackling a stimulating problem in physics. The Göttingen theatre was hopelessly provincial and this film of Brecht's "Dreigroschenoper" impressed them all tremendously.

After his car accident Landau took a long time to recover from the resulting amnesia. He had particular trouble with personal matters. He could forget, for example, the name of the doctor who was treating him.

"Good morning, Lev Davydovich, do you recognize me? Do you remember my name?

"No, I don't. But do you remember the Threepenny Opera? Here, listen:

> Und nun kommt zum guten Ende
> Alles unter einem Hut.
> Ist das nöt'ge Geld vorhanden
> Wird das Ende meistens gut!"

And so on, until the end of this latest brief spell of lucidity.

All Born's young people sang the songs from this film. They coined new versions of Mack the Knife's song to fit all sorts of occasions. The one about Pauli was particularly good.

This was indeed a single household – a cheerful and kindly family. They arranged frequent parties as well as country walks both short and long. And, of course, there were endless discussions which often went on far into the night and even until the next day dawned. If they found they had talked themselves into a dead end they would go to Max Born who was glad to discuss ideas and papers from any field of theoretical physics or mathematics.

Each could pursue whatever question he or she liked and Born was never taken by surprise, no matter what his students and assistant said or showed to him. Born never imposed his own thoughts or tastes on anyone. If a thought came to him he would tell everyone about it and it would be taken up by whoever it happened to interest at that particular moment. Rumer, for instance, worked as Born's assistant in the field of quantum electrodynamics but he did not abandon the general theory of relativity and still pursued topics in pure mathematics. He obtained some very important results in a completely new sphere – quantum chemistry.

In 1927 Heitler and Fritz London had published a paper in which for the first time the energies of the hydrogen molecule were calculated using a quantum approach. Rumer now took up this matter. By 1930 the "Göttingener Nachrichten" would have published a joint work by Heitler and Rumer – "The Quantum Chemistry of the Multi-Atom Molecule". This paper, along with the others of Rumer's Göttingen period, was to be a seminal work in the new-born science of quantum chemistry.

"Göttingen had nothing to offer except its fame and its brilliant professors," Yuri Borisovich used to say, "and if you really wanted to learn then you could learn. If you didn't want to learn then nobody would teach you. Some could cope with this and some could not.

Robert Oppenheimer, for instance, was one of those who managed. He was rich but he closed his bank account, took no money from his parents and lived in Göttingen like everyone else. He worked and studied just as we all did. He did not do anything out of the ordinary at Göttingen but he did learn quantum mechanics.

On one occasion Oppenheimer came to Max Born with a complaint: why had Rumer been given an award while he had not? He just could not grasp the idea that Rumer was poor whereas he was rich. He still went around as one wronged, asking: "What has Rumer done that I haven't done? And what is an award anyway – aid to the poor or a recognition of achievement?"

But no one among them took such trifles seriously – there were no really deeply hurt feelings – they were young and open-hearted and they trusted each other completely. If anyone had a real problem or any kind of misunderstanding arose everyone tried to help – and Born was always there to advise them.

The workers of a local factory which mainly turned out school microscopes and other fairly simple optical instruments once invited Rumer to come and speak to them about Russia. They baked a beautiful cherry pie and arranged a tea-party in honour of their Russian guest. Rumer had plenty to tell the workers and at the end of his talk they bid him a very warm farewell, excited and surprised by what he had told them. The next day Rumer received a summons to report to the police. Somewhat perturbed, he went to Born and told him everything. Born was upset:

"What a stupid thing to do! Why did you go? They could deport you as an undesirable alien. They'll say: 'What did you come here for? To study and work, or to spread propaganda? They won't do anything to you, of course, but nevertheless. Let's ask the advice of Frau Born. She has had a lot of dealings with the police."

This last remark was, of course, a joke. It was an allusion to the occasion when Frau Born went for a ride in a motor-car. It was at the time when those who could afford them bought cars and drove them around the crooked streets of Göttingen with scant regard for any rules or regulations. The Borns, too, bought a car. Born got into the driver's seat, drove a few metres across the yard and declared that this business was not for him – that his driving days were already over. Frau Born, however, decided not only that she would learn to drive but also that she would study the inmost workings of the machine itself.

On one of her first excursions Frau Born drove too fast round a bend and ended up on someone's lawn. She did not manage to get off the lawn for there she was apprehended. She was taken to a police station and charged with exceeding the speed limit in a built-up area. The punishment for this offence at that time was either three days detention or an on-the-spot fine of 15 Marks. The senior police officer wrote out a slip for Frau Born and said:

"Go and pay the 15 Marks please, Frau Professor."

"I'm not going anywhere. You say it's either 15 Marks or three days in detention. All right, detain me."

"Frau Professor, what are you saying? How can we do that?"

She argued with him for a long time, trying to convince him that she wanted to find out what it was like to spend three days in a cell – but she could not shift him.

And so it was that Frau Born was now called upon to advise in Rumer's brush with the police. She listened to the whole story with interest and said:

"Go there, show the summons and deny everything (verneinen Sie alles). 'Nothing of the sort happened, it's a pack of lies. I came here to study science under a German professor in world-famous Göttingen and you think God knows what about me.'"

Rumer did just that. He went to the police station and showed the summons. The police superintendent reproached him: "Herr Rumer, you are spreading subversive propaganda and at a time when every German should be putting his Fatherland first!"

Rumer, with a hint of suppressed agitation replied: "Sir, what do you take me for?" He continued as Frau Born had instructed, denying everything so convincingly that the superintendent started to apologise. He knew for a fact that Rumer had been at the workers' meeting, had enthusiastically sung the praises of his Socialist homeland and all that was being done there. He had spun such yarns – maybe he had exaggerated so much that nobody had believed him? In any case the superintendent let Rumer depart in peace and bid him farewell by saying: "Next time you give a talk about the Soviets send me an invitation."

Rumer, well pleased with himself and in high spirits, went back to Born to tell him how well everything had gone and how wise Frau Born's advice had turned out to be. He found Born in a very downcast state. As always when he was worried he had a headache. Just recently he had written several letters in an attempt to obtain a grant for Rumer – and now this business with the police. He was afraid that Rumer, the enthusiast might even try to win over the police superintendent to the Soviet point of view.

The only way to make Born forget his anxieties was to get him involved in a scientific discussion. Fortunately, this was not difficult to achieve. Rumer started to explain his latest ideas as if nothing had happened. He talked about the possible application of quantum mechanics in chemistry. In chemistry? Born was transfigured in an instant. He was, of course, already accustomed to hearing all kinds of unorthodox thoughts expressed by this young man from Russia who had so quickly been accepted in Göttingen. Mathematics, yes – but chemistry? This was completely unexpected! Born realized that Heitler's influence could be traced in this suggestion. In a rather delicate state, his head splitting, he began to listen to Rumer. The talk did not start well. Born sat at his table showing no interest whatever. Rumer sat opposite him in an armchair. Sitting there and making no attempt to get up and go to the blackboard, Rumer explained his new ideas, his main purpose being only to distract Born from his unpleasant thoughts. Finally Born said:

"Have you spoken about this to Heitler?"

"Yes, Herr Professor. He and I did the calculations together."

"So you've already done some calculations?"

"Yes – in fact, the final ones."

"Mensch! You should have said that right at the beginning! Erst losrechnen, dann nachdenken! (First do the calculating, then work out the implications – this was Born's favourite saying.)

That did it. Since the idea had reached the calculation stage Born was interested and wanted a full account. Soon, as Rumer came to the nub of the argument, Born broke in with his well-known metaphor: "This is where the frog jumps into the water!" That was it, the work was approved. Born gave his blessing to all Rumer's projects but nevertheless considered that Rumer's work better complemented Einstein's than it did his own. Meanwhile, Einstein was waiting for Rumer – but the young man from Russia was for some reason, in no hurry to get to Berlin. Rumer stayed and worked with Born.

Born, for his part, went on trying to obtain financial support for Rumer – and with some success. In a letter to Einstein on 22 February 1931 he wrote: "My colleague Rumer, about whose grants I have written to you before now, can stay with me for another year. My assistant Heitler is off to America in the summer (Columbia, Ohio) and Rumer will stand in for him – and I have managed to get hold of some money for the winter." [2, p. 17]

Born obtained money for his young co-workers from the Lorentz Fund, from the Rockefeller Foundation and from other such bodies set up to support talented young scientists. All the money came to Born and he apportioned it himself. Born also obtained funds from businessmen – from bankers, big industrialists – and from wherever else he could. He wrote detailed letters to these sponsors and visited them, too, giving introductory talks on modern physics.

There was, for example, one such benefactor – the owner of a stud farm in Frankfurt – who regularly sent Born money. Born, just as regularly, had to meet him, dine with him in his home and patiently explain to him what his friend Einstein's theory of relativity was all about. On every such occasion and with every one of his lectures at the "Gentlemen's Club", where not one of his listeners understood him, Born secured the "ransom" of one or another of his young scientists, who would be granted, as a result of Born's efforts, the right to exist.

For example, before Born succeeded in obtaining a grant for Rumer from the Lorentz Fund he got money for him from the Baron von Weinberg. This was that same "award" which so upset Robert Oppenheimer. Baron von Weinberg was one of the richest men in the German chemical industry. When Heitler and Rumer published their first paper on quantum chemistry Born immediately sent an abstract of this work to the baron. In his covering letter he wrote about the importance of the results which had been obtained and about the progress in the chemical industry which this work could bring about. Naturally, Born did not point out that this was a pioneering project, that quantum chemistry was barely two years old and that uncharted waters still lay ahead. However, he did say the following: "... As you will see from the text, there is a connection between this work and the work you yourself did in your youth, when you were still a student. I am pleased that my young colleagues here in the institute have been able to carry on where you left off." Some money arrived very soon afterwards. Thus Born performed his magic. There was even a club, the "Friends of Professor Born". Subscriptions varied – some contributed five hundred, some a thousand Marks, others very much less.

Science was then the private concern of a few. The time was still far off when science would become a complex system, a hierarchy, a matter in which governments would be involved, when institutes employing thousands of people would be opened or closed by the stroke of a pen. At that time young people were accepted into the world of science as into a circle of friends or a family. And for their sake Born wrote to Einstein.

Niels Bohr

"Then, As Now, Physicists Went About Their Work and Regarded It As Perfectly Normal"

Yuri Borisovich Rumer had been taken into "the bosom of the family" and accepted as a friend. One after another, memorable moments sparkled like stars.

The first time this metaphor occurred to Rumer was when he spoke to Einstein and Ehrenfest in that attic room. Ehrenfest had gone off somewhere. For some reason there was a violin maker in the house. This man came into the room where Rumer was waiting and struck up a long and involved conversation about the repair of violins. That was the first time that Rumer caught himself thinking: "These are the star-minutes of my life." This image was to return to him more than once.

After that Rumer had cheerfully accepted Frau Einstein's invitation to stay for dinner but Ehrenfest had poured cold water on that idea ("No, it's time to go ..."), leaving Rumer no choice but to decline the offer: "I have already been honoured here, I'll eat somewhere else."

"In the end I did stay for dinner after all with Einstein, his wife and Ehrenfest," wrote Yuri Borisovich in his recollections of his meetings with Einstein, "During the meal Einstein talked to Ehrenfest about the axiomatics of electromagnetic fields. Ehrenfest expressed a desire to meet my wife. 'I want to know if she will let you get on with your work, he said.'" [18, p. 110]

As they said their farewells Ehrenfest slapped Rumer on the shoulder, smiled his usual wide smile which completely transformed his face, and said in Russian: "Come to the conference at the university tomorrow. I'll introduce you to a young chap from Russia."

Ehrenfest pronounced the Russian word for "young chap" fault-lessly.

Professor Rumer's "Recollections of L. D. Landau" were published in the journal "Science and Life" in 1974:

"In these notes I do not want to comment on Landau's scientific works. Modern theoretical physics is not accessible to non-specialists and the ability to popularize this science is a special talent given only to a few. I do not consider that I possess such a gift myself even though Lev Davidovich Landau and I were co-authors of the popular book 'What is the Relativity Theory?'

Likewise I would not wish to give the slightest support to that folk myth in which Landau appears as a figure in "sandals and checked shirt". My reason is (and here I use a term which is most appropriate in Landau's case) that Landau's real centre of gravity is not here – not in his paradoxical utterances which make him the hero of so many anecdotes – but in the fact that he was a great physicist of world stature and the creator of an outstanding school of Soviet physicists.

Pavel Sigismundovich Ehrenfest introduced me to Landau at a conference on theoretical physics in Berlin at the very end of 1929. Landau said to me regretfully: 'Just as all the pretty girls are all spoken for and married, so are all the best problems already solved ... I doubt if I shall find anything worthwhile among those that remain.'

But he did find something.

In January 1930, while staying with Pauli in Zurich, he discovered what he called the last of the great problems – the analysis in quantum terms of the movement of electrons in a constant magnetic field. He solved this problem in the spring of that year while in Cambridge with Rutherford. So it was that Landau's diamagnetism and Pauli's paramagnetism came upon the scene at the same moment in the history of physics ...

It is said of Landau's character in his youthful years that he was cocky and outspoken in his judgements to the point of deliberate eccentricity. These traits reminded me of the young Mayakovsky when he used to go around in a yellow jumper and startle those who chanced to listen to him with his pronouncements about himself and his own significance. The resemblance demands an explanation. I think that such exaltation of the ego is characteristic of geniuses who are still seeking a position which is worthy of their talents. When Mayakovsky won general recognition he softened, became more approachable and kinder. Landau followed the same pattern. When universal recognition came to him – both at home and abroad – he stopped being arrogant." [34]

There are many stories about Landau's arrogance. His friends, pupils and pupils' pupils hand on these legends from mouth to mouth and in them there clearly appears an inevitable element of fabrication. We shall not "enrich" the distortions which, unfortunately, already exist in the literature about Landau. There are, however, also well-documented facts.

A typical example is Landau's telegram to Niels Bohr. In 1931, when Landau was in England, Dirac gave an account at a seminar of the work which had led to his famous equation. He tried to make sense of it in terms of "the positively charged electron" (the future positron!). Bohr held Landau's work in high estimation and sent him Dirac's paper, asking him his opinion. He soon received a telegram from Landau in which he gave his assessment of Dirac's ideas. The telegram was short and unambiguous: "Quatsch" – i. e., rubbish. This is yet another proof of the difficulty with which even brilliant minds bring themselves to accept new ideas. But no one in that age of heated argument took offence at such attacks.

Recalling that meeting at the conference in Berlin Yuri Borisovich said: "On that day of our first meeting we sat on the very top row of the lecture theatre. Einstein was addressing the meeting. Landau frowned and fidgeted constantly: "Oh! What nonsense! No, it's all wrong ... Yuri, do you hear what he's saying? Let's go down there, I'm dying to talk the old man (Einstein was then 50 years old) out of his unified field theory!" During the break I was horrified to see this mere stripling making his way down to Einstein. But no one would have been bold enough to break through the barrier of Majesty which stood in front of those who occupied the front row. No one would have had that kind of nerve, not even Dau. He went up and took a closer look at Einstein but stayed where he was – at a distance."

We know that Einstein's battle with Bohr and Born over the interpretation of quantum mechanics went on right up to Einstein's death. It was the younger generation which embraced the quantum faith without qualms. When you look at the bibliography of those years you are struck by the amazing frequency of great discoveries and the abundance of new names that appear alongside those which were already well-known. These names, then unknown, have now become so familiar that there is hardly a student struggling to learn the basic equations, rules and effects which now bear these names who realizes that these same equations, rules and effects were the work of very young newcomers to physics and the fruit of desperate conjecture, faith and doubt.

What a storm that was! What emotions there were then when these absolutely radical ideas about nature were just being articulated. "But we who came to work with Born at that time were not fully aware of the enormity of the changes that were taking place. We felt that all this was revolutionary but we did not feel this very keenly. Perhaps we lacked historical perspective. Then, as now, physicists went about their work and regarded it as perfectly normal."

Viktor Weisskopf (admitted to the University of Vienna 1926, worked with Born in Göttingen 1928–31) wrote with regret: "It was already too late when I came to physics; all the new fields were discovered. What was left for me to do?" [35, p. 20]

What Viktor Weisskopf did and what he achieved is well known to the world of science. But the feeling at the time was that "physicists went about their work and regarded it as normal."

Born's seminars. Cakes were bought, coffee was served in tiny cups. On the front row next to the professor sat his assistants and guests. Then came those studying for doctorates, people from all over the world. Everything seemed perfectly ordinary, sometimes even boring – especially at a seminar where Dirac was the main speaker. He was not much good at delivering papers.

Once when Professor Rumer was talking to some students at Novosibirsk State University one young theoretician said: "I found Landau's "Quantum Mechanics" excruciating. But Dirac – that was like reading the Song of Songs! Everything logically structured, clear and refined. What a pleasure it must have been to listen to Dirac himself!"

Yuri Borisovich surprised these young people by replying: "No pleasure at all, actually. Listening to Dirac was a dreadful experience. We accepted Dirac's ideas and fell under their influence only when we read his papers. But at a seminar ... Dirac would come in, no smile, no enthusiasm. He would take a piece of chalk in his long fingers and start writing formulae on the board, without saying a word. After a while Born couldn't stand it: "Paul, tell us, what are you writing?" Dirac, still writing, reluctantly started to speak: "W minus alpha ar pi ar minus alpha zero m c and all this multiplied by psi, then alpha mu by alpha nu ..." and so on in the same vein – and he sincerely believed that he was explaining.

I once went for a walk with Born and Dirac in the country just outside Göttingen, near Plesse castle. We talked about everything on earth, architecture, music, etc, and finally came to science. Born was excited about Dirac's theory of the relativistic electron and his very recent conjecture about the existence of the positively charged electron. Dirac shrugged and said nothing. "What will you find to delight us with next, Dirac?" Born asked him. Dirac answered: "Nothing more, nothing more." What else could be demanded of the man who wrote the equation which "explains the greater part of physics and the whole of chemistry"?

Born's seminars were at their most vivid when Niels Bohr, Ehrenfest or Pauli came to Göttingen. When Pauli came there were always desperate arguments and invariably some kind of hilarious happening. His ability to spot the weak points in a theory was legendary and it was very difficult to argue with him. In fact, it was considered that Pauli was always right. His horrendously inexorable comments would bring the argument to an end and, so that battle could not be rejoined, he could, without changing his expression, go on to tell a dirty joke in the same raised voice as in the argument. Niels

Bohr said of him: "Every one of us is afraid of Pauli. Actually, it is not so much that we are afraid of him as that we do not dare to admit to ourselves that we are not afraid of him." [11, p. 100]

When Niels Bohr came he usually gave a course of lectures. In fact the Bohr festivals became quite a tradition in Göttingen. They were real feasts in the scientific calendar.

With the arrival of Ehrenfest the very atmosphere of the place changed. Relations between people took on that vivacity and directness which was a little lacking in Born's strict and rather stiff communication with his students.

Yuri Borisovich used to say: "There was no limit to what Born did for me. He did as much as any one person can do for another. But the relationship between us varied a lot. Things were not always easy."

Everything depended on what mood Born was in and that was made obvious immediately. For example, if he greeted Rumer with: "Guten Tag, Doktor" it was clear that he was in a bad mood. That meant: I've been sitting all day scribbling on bits of paper, screwing them up and throwing them into the bin. Nothing came of it. I don't even want to talk about science. You can take your leave and go and get on with your own work or go to the cinema.

If, on the other hand the greeting was: "Guten Tag, Rumer", that meant that his mood was somewhat more cheerful. Something had suddenly become clear and there was a topic to be discussed – not connected with your work but something that Born was working on himself at that moment.

And then there were times when Rumer was addressed as "Lieber Rumer!" That was the signal that Born's work was going swimmingly and that he would now set his own papers aside for a while and you had time to give an unhurried account of your own problems.

With Ehrenfest, on the other hand, one could discuss anything at all, even the most abstruse ideas, at any time. He was never dismissive about anything. He was considered to be the conductor of the European scientific orchestra – but he was also the teacher of harmony. He did not acknowledge any struggle for supremacy and would not stand for any talk about someone's ideas being stolen by someone else. Steal an idea, what does that mean? Could you steal an idea from Einstein, or Bohr, or Rutherford?

At the seventh Solvay congress in October 1933 Paul Langevin in his opening speech said of Ehrenfest: "Many of those present here were his pupils and all were his friends ... Ehrenfest, who did so much to build a bridge between the old ideas and the new, was in many ways the soul of these gatherings . Better than any of us, he knew what difficulties we face. I was looking forward joyfully to meeting him here, hoping to experience once again that inspiring influence which Ehrenfest radiated when he spoke to his students and friends. However, we shall not see him again. The crisis in our science was embodied in him and in the tragedy which engulfed a great mind and a great heart." [36, p. 160]. Paul Ehrenfest had committed suicide on 25 September 1933.

Born's weekly seminars were held on Wednesdays. The topic was not, as a rule, announced in advance but everyone knew that it would be something important and would make an impact. Before the start of the seminar Born might ask the gathering to wait for 10 minutes, saying that he needed to think for a moment. Then ten minutes later he would appear with Ehrenfest, who had just arrived. Those present would show their excitement by stamping their feet. This is an old student custom in Germany – stamping instead of clapping.

Born and Ehrenfest always exchanged a few caustic comments. Born would be giving his talk, explaining his latest work, naturally, in terms of matrix mechanics. Ehrenfest would interrupt him:

"Max, why do you keep on piling up these matrices one on top of the other? Write up Schrödinger's equation and in two minutes everything will be clear."

Born, shrugging his shoulders, would reply: "Each to his own. I'm used to matrices. They suit me better."

"But, Max, surely you're aware that there are good habits and bad habits. One should try to shake off the bad ones."

There was a very free and easy atmosphere. At the end of the seminar, for example, Heitler might raise his hand and say that he had an important announcement to make. No one thought anything was amiss on the occasion when Heitler came out to the front and said: "Nordheim and I have discovered a surprising correlation between the radius of the Universe according to Eddington's theory and the mass of the electron." Everyone started to snigger, expecting some kind of spoof, but noticing that Born was taking it seriously ("Oh yes, tell us about it.") Heitler went on: "Herr Professor, as yet we cannot present this in full, in the form you require with your Göttingen precision. For now we shall just have to present it as it is."

It very soon became clear that this had all been dreamed up to entertain the meeting. It was greeted with appreciative stamping. Born frowned.

"You should not be angry, Professor. We sent this paper to Ehrenfest in Leiden and we received this reply. Do you mind if I read it out?"

Born did not object. Heitler pulled the letter from an envelope bearing a Leiden postmark and read: "I congratulate you on these remarkable ideas. They are definitely worthy of attention. Do not be concerned if progress is slow at first. I will look after Born."

Max Born was not particularly fond of being teased but the young people trained him to go along with their tricks. With some of them he was actually delighted. On one occasion his students composed an announcement and had it printed in a Hanover newspaper. The public was informed that at 12 noon on such and such a date, as part of a research programme, the James Franck Institute of Physics would be releasing some balloons which would drift over various parts of the former Kingdom of Hanover. Anyone noting the descent of any of these balloons in any area was asked to notify the physics institutes without fail. If any balloon was seen passing overhead, then the time and the direction in which the balloon was travelling should be noted.

The observer should pass on this information and include, if possible, details of his or her trade or profession. The James Franck Institute was inundated with letters. It turned out that an enormous number of people, especially priests, had seen these non-existent balloons.

James Franck and his young people were regular attenders at Born's seminars – and Born's students often got themselves invited to Franck's laboratory to try their hand at experimental work. If Jakob Frenkel or Walter Bothe came to Born's seminar they were invariably taken to the Franck Institute to look at the new apparatus, to discuss experimental results on the spot and to twiddle the knobs on the instruments. Likewise, if Gustav Hertz or Charles Wilson visited James Franck they would never omit to visit Born's institute for a chat with the theoreticians.

In honour of such guests Born, as a rule, arranged parties at his own home and never forgot to invite his assistants. Yuri Borisovich gave an account of one such occasion which particularly stuck in his memory. This was the party held in honour of Rutherford. During the day Rutherford gave a lecture to a packed "Aula" about problems facing contemporary physics. The lecture was delivered in English but Rutherford's exceptionally clear New Zealand accent made it comprehensible to all. After the lecture came a ceremony by which the honorary title of "Doctor of Georgia Augusta" was conferred on Rutherford. In his speech of acceptance, after the gown had been put on his shoulders, Rutherford said: "I believe that science is simple because I am a simple man."

That same evening Born's guests gathered at his home. Among them were James Franck and Richard Courant with their wives, Born's three assistants – Heitler, Nordheim and Rumer and, following the official ceremony, a high-ranking official, the administrator of Georgia Augusta.

"When we went into the drawing room," related Yuri Borisovich, "the important administrator gave each of us two fingers to shake and peered at us with an expression of disdain. The great Rutherford, however, reached out his huge hand, took our hands warmly in his wide grasp and shook until it hurt. Then he turned to Born: 'So these are the assistants you've got now. The last time I came, I remember, you had Heisenberg!'"

For a while at dinner the Göttingen professors were on their best behaviour and the young people sat in total silence. The expression of disdain remained on the face of the important administrator; he was apparently shocked by Rutherford's homely manners, loud guffaws, shoulder-slapping and artless comments. Rutherford ate with gusto, resorting only infrequently to the use of knife or fork. When the dessert arrived everyone heard distinctly the loud crunch of choux pastry; the cakes were clearly to Rutherford's liking. Satisfied at last, Rutherford lit his pipe. Then something unexpected happened.

By this time the conversation had become less formal and, as often happens among middle-aged people, had turned to reminiscences. James Franck recounted an episode from his war service. He had been on guard duty one day in front of his unit's staff H. Q. Suddenly he saw a magnificent officer on horseback thundering towards him. The officer reined in the horse inches away from the somewhat shaken Franck.

He slid to the ground, threw the reins to Franck and said: "Hold these, I won't be long." Franck was left looking after the horse rather longer than he would have wished. At last the officer emerged, took out his purse and offered Franck 50 Pfennigs, saying: "Sorry, soldier, for keeping you so long". "I have to tell you", answered Franck, "that in view of my profession and my position in civilian life I am not able to accept tips." The officer looked at him closely, smiled and said: "Never mind, soldier, don't be shy. Money is money, no matter what. Even in civilian life you'll find a use for it."

"Well, did you take it?" asked Rutherford.

"Yes."

"That was a mistake. You should have asked for more."

The important administrator winced.

James Franck's story reminded Richard Courant of his own days in the army. He told how he had been in action, had been wounded in the stomach and had been decorated with the Iron Cross. During the terrible social unrest which followed the war and later, when the newly-elected national assembly met at Weimar and the constitution of the republic was put together, Courant was active in politics. He was the chairman of the Soviet of soliders' deputies for the town of Göttingen and the surrounding districts.

"This was the sort of thing I did. I would write out, a chit for example: 'Citizen so-and-so is allowed to proceed to such-and-such a place and return for the purpose of acquiring five bushels of potatoes' – than my signature: chairman of the Soviet of soldiers' deputies, Doktor Courant."

Rutherford was amused: "Doktor Courant, I know now why nothing came of your revolution. You spent too much time thinking about mathematics and not enough about your coup d'état."

The university administrator could not take any more. He rose ostentatiously, said goodbye only to Born and left.

After that began the musical entertainment without which no party at Born's house was complete. Born loved to play the piano and was always the first to perform at his own parties. He always played some little-known piece. Then Hedi Born would take her place at the piano. She played superbly, usually selecting Mozart or Chopin when she chose Schumann, singing one of his songs to her own accompaniment. This occasion followed the usual pattern. When Frau Born had finished playing Rutherford applauded loudly, then confessed from the heart: "Thank you, Frau Born, that was lovely. I must admit I do not know much about music but I thought you played very well. Now when Professor Born was playing, it sounded as if he was working out problems in algebra."

To round off the evening they all went down to the central square. Rutherford was required to climb up on to the fountain and, like every other newly-fledged doctor of Göttingen University, kiss the little shepherdess. This Rutherford did with great pleasure. There was just one moment of alarm when he climbed over the barrier surrounding the fountain. Although he was 60 years old Rutherford was still immensely strong and the young people feared he might break the old cast-iron railings.

Chapter 9

Metamorphosis

Göttingen was now at the zenith of its scientific fame. "With more justification than ever it could now be said that, in this quiet little town with its avenues of limes and its respectable houses solidly built in that 'Jugendstil' which was already out of fashion, the international community of mathematicians now held a never-ending congress. Numerous scientific complexes and laboratories ringed the town, as if replacing the medieval wall. The Mathematics Institute had settled into its new building. The 'Lesezimmer', had grown into a large, well-lit library... In sunny weather students and their professors could be seen sitting at the tables of street cafes discussing politics, love and science. The little shepherdess still gazed serenely into her fountain ... Outside Göttingen there was no life." [26, p. 248]

With equal justification it could be affirmed that an international congress of physicists was also in permanent session in Göttingen. To come to Göttingen and give an account of one's work at one of Born's or James Franck's seminars was either to win complete recognition or to be discredited, depending on the result. This was now the accepted test.

No physicists from Göttingen attended the sixth Solvay congress in 1930. Straight after it Sommerfeld travelled to Göttingen to present his paper on "Magnetism and Spectroscopy", with which he had opened the congress. It was important to him to know how the Göttingen physicists assessed his work. During Sommerfeld's visit a minor incident occurred which Rumer witnessed. When he reported it afterwards opinions were divided: some decided that it must have been a joke, others thought it might have been meant seriously. After the seminar Sommerfeld had asked Born to show him the library. Born had given Rumer the job of accompanying Sommerfeld to the "Lesezimmer".

Sommerfeld was a little man and his hair was already completely grey but with his protruding black whiskers and military bearing he looked like a typical Prussian officer. As they walked together, Rumer, towering a good 30 centimetres above him, felt clumsy and huge. They entered the library and Rumer went over to the big table where the latest journals lay. Sommerfeld, however, showed no interest in them whatever. He made straight for the shelves. There he found the "S" shelf, counted the books by Sommerfeld standing there and declared with satisfaction: "Not a bad library".

Less favourable opinions were also expressed. Pauli, for example, could come to Göttingen, speak to no one, listen to no one and then leave a note on Born's desk: "Was in Göttingen. Cakes and beer marvellous. Physics, as always, dismal." There was also an occasion when young Janos (John) von Neumann, who was already a professor at Princeton, came to Germany and travelled around by car with a friend. Somewhere in the Harz the car broke down. Neumann, not wishing to waste time, left his friend and a mechanic to carry out the repairs while he set off for Göttingen alone. In Göttingen he called on Born to discuss quantum mechanics, went on to talk to Courant about differential equations and then visited the astronomer Heckmann to talk about stars.

He immediately spotted the weak points in Heckmann's theory of the evolution of stars and made the suggestions which, in a more developed form, were to be incorporated in the work of Ambartsumian a few years later. When the car was ready Neumann went back to the Harz to carry on with his tour. From various points along his way he sent letters to the people he had spoken to in

Göttingen, answering the questions which had arisen but had not been resolved in their conversations.

"Outside Göttingen there was no life!" In fact two kinds of life were lived in Göttingen. An invisible wall separated two kinds of people: those who lived the life of the mind and stood by moral principles and those who were far removed from any abstract knowledge, who recognised only common sense made manifest in worldly well-being. The symbiosis between these two types was no bad thing in those days before the town, along with the whole of Germany, came to be faced with the most dreadful ordeal in its history. For the moment, the two strata lived together peaceably, each helping the other, maintaining contact where necessary.

Then, one day, a Göttingen newspaper announced that perjury had been committed in a Göttingen court! Other newspapers soon took up the same story. In Germany at that time perjury was as grave an offence as murder. If anyone was found to have lied under oath then the normal court was adjourned and a special court – the Court of Jurors – was convened. This court was made up of three judges and six lay jurors – the so-called "Schöffen". The main function of this court was to examine charges of two categories: murder (Mord) and perjury (Meineid).

The owner of a small baker's shop, a man already getting on in years, had married a young girl. She was the only daughter of a moderately wealthy citizen who had lived next door to the baker. The girl had grown up without a mother and when her father died, leaving her a modest inheritance, she had agreed rather hastily to marry the neighbour. As time went on, however, the young woman had lost interest in her elderly husband and had found herself a young man.

The pair started to meet secretly. The husband soon discovered what was going on and put in an application for divorce on the grounds of his wife's infidelity. He asked to be allowed to retain all the money brought to him by his faithless wife when they married. Neither the divorce nor the loss of her inheritance suited the young woman and she begged her young lover not to confess in court to their secret liaison. She assured him that she, too, would deny all.

The young man kept his word. The young woman did not keep hers. She had already raised her hand to take the oath but then lowered it and said: "I admit that I intended to swear that I am innocent, but I am guilty." The young man, who had already sworn that there was no relationship between them, was exposed as a perjurer. The baker forgave his wife. The case was dismissed and a new trial began. The "Schöffen" court was convened to try the young man with all due rigour.

All sorts of people attended this trial. One of the three judges happened to be an acquaintance of Born's. He had been an officer in the German army in the First World War and had been taken prisoner by the British. Captured officers enjoyed special privileges then. So that they should not waste their time they were given the chance to occupy themselves with various activities. They could even learn a new profession. This particular man had studied jurisprudence, then had taken the assessors' examination and become a barrister. It was fairly obvious that he was a Communist sympathizer, although he never openly said so.

During the trial of the unfortunate young man he broke the rules by entering the jurors' chamber to find out how unanimous they were on the need to impose the ultimate penalty. Their unanimity was complete. "I looked into the eyes of each of these jurors in turn", he told Born, "and saw what a collection of self-satisfied burghers they were. I asked them: "Are you fully aware of the fact that you are sending to the gallows an irreproachably honest man whose only crime was to defend a woman's honour?" I saw before me only a row of stupid faces regarding me with contempt."

The court sentenced the young man to 12 years' imprisonment. At this the academics and foreigners living in Göttingen tried to intervene. The same judge, Born's acquaintance, took no part in the campaign personally, did not write to President Hindenburg, etc, but at each step told Born, Franck and the others what they should do. Born went off on missions to various places, as did Franck and Courant. From some of these journeys they returned encouraged, from others despondent. Finally someone came back and announced that the President had pardoned the young man.

President Hindenburg was still in power but alarming signs, warnings of the storm to come, were already to be seen. Yuri Borisovich remembered how the change first became evident in the town:

"In this small town where, it seemed, everyone knew everyone else, strange things started to happen which we could not quite understand or did not try to understand. The people of the town with whom one constantly had dealings – such as shop-keepers, the laundry proprietor, the minor officials – were drawn more and more to the ideas of Adolf Hitler. No one wore any badges at first. Then these started to appear. No swastikas in the beginning, just Iron Crosses. The Iron Cross was brought out initially as a badge of courage. It simply announced that the wearer had taken part in the war, had been found worthy of decoration and was now getting on with his normal life.

Then the Iron Crosses were exchanged for swastika medallions. Even then many hid them shiftily under their lapels or placed them in such a way as to make them hardly noticeable. Near my house was a grocer's shop to which I went in all innocence to buy cheese and other provisions. The proprietor watched me jealously lest I should patronize other establishments. I once walked past and wandered into a different shop. In the evening this shopkeeper ran over to our flat in a state of agitation:

'Herr Doktor, today you went to a different shop. What is it, are you not satisfied with my cheese?'

'Nothing of the sort,' I replied, 'I was just lost in thought. I didn't realize I had passed your shop. It just happened that way.'

'Herr Doktor, you mustn't think that I am a Nazi.'

'Of course I don't. Why would a sensible fellow like you get mixed up in such nonsense!?'

Nevertheless, this man shamefacedly kept a swastika in his pocket. Only now and then, when he was quite sure that the person who had come in to buy his cheese was an Aryan, did he pin it to his chest. He quickly removed it if he thought a customer of Jewish extraction was entering his shop. He bowed to both with equal courtesy. In fact in rather too servile a manner for my taste: 'Many thanks for not forgetting me, it is a pleasure to serve such a fine young man.'

These swastikas suddenly began to appear in more and more buttonholes. People stopped hiding them in their pockets but still, on meeting a person whose nationality was not altogether certain, would, just in case, press hand to heart, thus hiding the swastika, and bow very politely. Later these badges became more firmly attached.

On one occasion I was out walking with Hannah Heckmann. We were enormously pleased with each other's company.

(Let me explain that Mila and I still lived together as good friends but the love which had grown up between us and united us had had its day. As she was a very beautiful woman whom many admired she soon had men friends. Neither of us made heavy weather of this. Hannah Heckmann was the wife of Heckmann, the then famous astronomer. My life was completely at the mercy of the heavens: if the stars shone brightly then Hannah's husband would be sitting in his observatory and Hannah would be mine for the evening; if the sky was overcast I had to make do with the quantum theory of chemical valency.)

Anyway, on this particular evening a young man walked up to us. I can still remember his face. He was a well-built fellow. I was struck by the hate in his eyes. He held out some kind of paper and thrust it into Hannah's hand. In Germany at that time it was quite normal to have a cinema programme given to you by a passing stranger. That was how they advertised new films. We assumed that this was just another film programme and, since we were not intending to go to the cinema, Hannah crumpled it up and would have thrown it away. I stopped her: 'Don't throw it away just yet. Let's have a look at what's on.' It was then that we saw the words printed on the sheet: "Girls who consort with Jews will be punished in the Third Reich." It did not take much to make Hannah laugh. She started chuckling. The young man looked at her in surprise and walked away.

No one believed that the Nazis could come to power. I once went to the post office, where I regularly picked up, each month, the sum of 25 Marks (this was the money that Baron Warburg continued to send me). The clerk, who knew me well enough by then, first got my money ready, then held it in his hand and demanded to see my passport. All right, I thought, if you want to see it have a gape at this – and I held out my enormous red placard of a passport. He was unpleasantly surprised and instantly displeased:

'Who issued this passport?'

Oh, I thought, so you're asking who issued this passport. 'The Moscow Soviet of Workers' Deputies,' I rapped out loudly and slowly.

'What do you mean?'

'Just what I say. In the Soviet Union passports are issued by the local Soviets.'

Very angrily he threw me the 25 Marks and made as if to slam his window shut but I managed to inform him in a loud voice that if Germany went along with Hitler instead of the Communists it would come to grief. This man was very soon wearing his swastika all the time although he was a state official and therefore forbidden to wear the badge of any party."

Rumer was once sitting in Born's room at the assistants' table, doing the calculations which Born had asked him to do. Suddenly a loud voice boomed in the entrance hall:

"Well, Max, what are you up to these days?"

In came Karmann. Theodor von Karmann, the great mathematician and mechanical engineer, a Hungarian by nationality, had been a close friend of Born's student days.

"Max, I have come to say goodbye to you. I am going to America. They are building an institute for me in California. They tell me the first bricks are already laid."

Born was puzzled. He was sure that Karmann must be joking. They had lived side by side for so many years. But Karmann was not joking.

"You're out of your mind!" Born said to him. "What are you going to do in that country which is alien to you in culture and everything else? How do you think you'll survive outside Germany?"

"Survive. Exactly. I'm leaving to save my skin. I shall not live through this revolution of theirs."

"You're just running away. What revolution are you talking about? All right, then, let's ask Rumer. He is our local expert on revolutions and can speak from personal experience. Let's see what he will advise you to do. What do you think about all this, Rumer?" "I think you're doing the right thing, Professor Karmann. This is not a revolution but a kind of general sickness of the mind. If it continues like this I shall soon go home myself."

"You've all gone mad", said Born. "Of course it won't continue! All this stupidity will soon be forgotten."

But it did continue, more and more violently. Yuri Borisovich was struck by the contrast between two torchlight processions which were held one after the other. The very fact that two such processions were organised in such quick succession was itself unusual. The torchlight procession, the "Fackelzug", was a tradition among German students. It might be held to celebrate the anniversary of a favourite professor, the arrival of a distinguished guest or some other such important occasion. These torchlight processions, however, which began as light-hearted celebrations, were being transformed before Rumer's very eyes into mass marches of storm-troopers.

The first torchlight procession which Yuri Borisovich witnessed was held at the very beginning of his stay in Göttingen to mark the arrival there of a Jesuit theologian from Munich, a certain Abbot Guardini.

Religious affiliation was once a major consideration in the selection of professors in German universities. As time went by religious background mattered less, except in theological circles where it remained an important criterion. In fact, in order to smooth out these antagonisms over professorial appointments a little it was decided that representatives of the two main religious groups should deliver lectures on an exchange basis.

Göttingen's Professor of Theology, a Protestant, went to Munich to give a lecture in which, as he announced in advance, he would present some new evidence concerning the history of the Near and Middle East in the time of Christ. The very fact that Christ was treated as an historical figure rather than as a divine person was sufficient indication of the alien nature of this approach as far as the Catholics were concerned. The latter affirmed the basic principle that their role was to carry out the revealed will of God here on Earth. The lecture to be given by the Göttingen theologian would explode like a bomb among the Munich Catholics.

As to the topic which the Catholic theologian from Munich intended to expound to his Göttingen audience, half of which would be made up of non-believers and foreigners, that was not made known beforehand.

All the youth of Göttingen went to the station to meet their guest from Munich. A rumour went around to the effect that the papal nuncio himself, Cardinal Pacelli, would be accompanying the Abbot. The young people, who had never seen a cardinal in their lives, were eager to get a glimpse of the magnificent reception that the local authorities were preparing for their distinguished guests.

The train from Munich arrived in the evening. The red carpet was rolled up to the third carriage. A man in a red mantle and a bright red cardinal's hat stepped out on to the platform, diamond rings flashing on his fingers. He walked unhurriedly, giving his blessing to all the Christians and non-Christians of cosmopolitan Göttingen.

This was indeed Cardinal Pacelli, who soon afterwards became Pope Pius XII and who was to be so indulgent towards the bestialities and atrocities of Fascism. Together with the Cardinal, walking slightly behind him, was a short man with shining black hair. This was Professor Guardini. Late in the evening of that same day the students held a noisy and cheerful "Fackelzug".

The next morning everyone gathered in the Aule where Guardini was to give his lecture. No one yet knew what it would be about. The Aule was crammed with people. Guardini came in and the whole audience stamped its feet in greeting. The Abbot raised his head and looked intently around the hall. There was silence.

"Let me tell you a story which shook me, which brought unease to my soul. It is the legend of the Grand Inquisitor from the novel by Dostoyevsky.

Guardini soon had his audience under his spell. He told his story in a somewhat theatrical manner speaking beautifully in the language of Goethe, not quite in modern German, and emphasizing certain points with economical but very expressive gestures.

"And for all these ages Mankind prayed with faith and ardour: "Lord, our God, reveal Thyself to us". For all these ages the sons of men cried out unto Him that in His boundless mercy He might descend to His faithful people". And the Abbot went on to relate how Jesus Christ in the form of a man visited His children, descending to Earth in that very place where the fires crackled without ceasing, consuming heretics "ad majorem gloriam Dei" – to the greater glory of God. He came down to the hot streets of Seville, the city where the Grand Inquisitor Torquemada sat in dread judgement and dispensed chastisement.

The Lord came silently to His people with the sun of love in His heart. A miracle: He was recognized everywhere; all, both old people and children, stretched out their hands to Him with hope and love and kissed the ground where He had passed. He stood in the porch of the cathedral of Seville as a small white open coffin was carried in. In it lay a seven-year-old girl, the only daughter of a nobleman of the city. "He will raise your child from the dead!" cried the crowd to the poor mother, who was mad with grief. She fell at His feet: "If thou art He, then raise my child!" And He in His compassion quietly uttered the words: "Talitha kumi" – "Arise, o maiden!" The girl rose up, her hands still holding her white roses. Overwhelmed, the people prostrated themselves, sobbing and crying out: "It is He!"

Meanwhile the Grand Inquisitor himself was in the square, watching all this from afar. His eyes flashed with a sinister fire and he ordered his guards to seize this man. And such was his power that the people could offer no resistance. They bowed their heads before the Grand Inquisitor and let him arrest Jesus Christ.

The day passed and the night came. The Grand Inquisitor came alone to Him in prison: "Art thou He? Art thou? Do not answer. Be silent. And what couldst Thou say? Thou hast no right to add a word to that which Thou hast said before. Why hast Thou come here to meddle?" Christ remained silent. "Judge us, if Thou canst and art so bold. Know this, that I fear Thee not. Again I say to Thee, tomorrow Thou shalt see this obedient flock rush at my signal and scoop up burning coals to light Thy pyre, on which I shall burn Thee for coming and hindering us. Tomorrow I shall burn Thee. Dixi – I have spoken."

"Dixi" – said Guardini and paused. For a long time he was silent. There was not a sound in the hall. Then, in great agitation, as if in fear of that theme which he himself had chosen, he went on: "And now I had to answer the most important question: What should Torquemada have done? A dreadful thought crept into my mind. I was afraid of that thought, I did not want it. I was frightened, too, by the question I had put to myself. The perplexity of spirit which gripped me was so strong that I threw myself on to my knees and cried out: "Lord, forgive Thy servant. Give me the strength to make the right decision." With these words Guardini sank on to his knees and bowed his head.

There was a breathless hush in the auditorium. The impenetrable face of the papal nuncio paled a little. The prayerful silence of the speaker himself became intolerable. Then he sprang to his feet and addressed his listeners:

"Yes, Torquemada was right. He had to slay our Lord, Jesus Christ!" – and with wide sweeps of his hand he crossed himself. There was not a sound in the hall. The papal nuncio rose calmly and made for the exit beckoning to the Abbot to follow him. The audience watched them leave in silence. Then, when they had gone, there was an explosion of unanimous indignation.

Now, however, a cruel transformation was taking place among these once like-minded people. A rumour went around Göttingen that James Franck was being offered a chair at the University of Berlin. The gloomy presentiment of all was that he might accept – but it was soon revealed that Franck had refused to go to Berlin and was staying in Göttingen. A torchlight procession was held. The various student societies turned out with their drums, flags and costumes. In front went the "Steel Helmets", followed by the "Swabians" and "Thuringians" and then by the students who were not members of any society. They hailed Franck, who stood on his balcony and bowed in all directions. They sang joyful songs and shouted triumphant slogans.

"On the very next day after the procession which honoured James Franck", said Yuri Borisovich, "another Fackelzug was held. This time it was motivated by dissatisfaction at the rejection of some bill by the Reichstag or something of the sort. On this second procession there was no singing – only threats and swastikas. The mood was that of a pogrom. And the faces were the same as the day before – that is what made it so terrifying."

Chapter 10

"I Shall Leave Neither Tomorrow Nor the Day After, But In a Week From Now."

The storm clouds were gathering over Germany. The black swastika on its white background had spread like the plague across the whole country. "The pain I felt was of a very particular kind," recalled Yuri Borisovich. "After all, I had been brought up by a German woman whom I had loved madly as a boy. She had read German folk tales to me and sung kindly little German songs. I loved her house with its pretty curtains, her canary and her old parrot. From the moment I arrived in Germany I wanted to go and see her. I found out that she was living in Berlin but on my first two trips to Berlin I did not manage to meet her. We finally met when I went to Berlin for the third time in the summer of 1931. Alisa Bleker was then 70 years old. I was 30. I was immensely glad to see her. She was already quite grey and shrivelled and she wept for joy: 'My dear little boy, my Yurochka.'"

Yuri diverted her attention from her tears by telling her about his successes, about Born, Heisenberg and Pauli. He told her he had been to see Einstein, knowing this would please her. He told her how he had been invited to give several lectures in the higher technical college in Hanover. Professor Fuss had once come from Hanover to see Born and had heard Rumer speaking at a seminar. Rumer's work on quantum chemistry interested Fuss. He was also working on related topics but, as he put it, he had not yet managed to infect his students with the quantum chemistry bug. He therefore invited Rumer to give some lectures and an account of his work in this area. When Rumer asked Born's permission to go, Born said: "Accept immediately. It won't do you any harm to earn a bit of money." It is even possible that Born had planted the idea of inviting Rumer in Fuss's mind.

Rumer went to Hanover, where he discovered that he could not give his lectures because he did not possess a black suit with jacket and waistcoat. A black three-piece suit was an indispensable requirement at that time for all who aspired to give lectures at the Higher Technical College. However, since Rumer did not have a standard figure, and since suits of non-standard sizes could not be hired, Professor Fuss went to the rector to ask him to allow Doctor Rumer to give his lectures in an ordinary suit. He promised a properly pressed blue suit and some very interesting lectures. The rector replied with great dignity: "Your Rumer might as well give his lecture in his underpants. Are you seriously suggesting that a man who stands up in a blue suit to give a lecture to a group of students can feel that he is a genuine expert on Schiller and Goethe?" Understandably, Professor Fuss could not think of an answer to this and so Rumer was obliged to pay three Marks for the hire of a three-piece suit. This was an excellent suit with two rows of cloth-covered buttons and a satin collar – but it was far from being a good fit. The trousers were too short. The legs were intended to be pulled down over the wearer's boots and had elastic bottoms to keep them in place there. The problem was that the elastic bottoms gripped better than the waistband and the trousers constantly threatened to fall down. The wide jacket hung on Rumer's thin shoulders as if on a coat-hanger. Rumer felt more like a music-hall compère than a "genuine expert on Schiller and Goethe".

His lectures went very well. It was decided that they should be properly typed up and duplicated. Two of Professor Fuss's students were given the job of transcription. During this process Rumer witnessed a scene which, as it turned out, he regretted describing to old Alisa.

Rumer was sitting in Fuss's study with these two students and checking that they were writing in the formulae correctly. The telephone rang and Fuss was told to release the students immediately as they were urgently needed by the National Socialist Party Committee. At this Fuss, confident in his authority as college professor, rang the Committee and demanded an explanation. Five minutes later a storm-trooper in an impressive uniform walked into the study without knocking and addressed Fuss quite insolently:

"Yes? I understand you wish to change the arrangements."

"Yes, young man. These students of mine have been given a task which is of some importance and I would like them to be able to finish it."

"They are required immediately by the Party to help in preparations for a parade."

"Allow me to suggest, young man, that the transcription of our guest's lectures, which must be completed before he leaves, is more important than a parade."

"?!" The storm-trooper did not honour Fuss with a reply.

"Could you possibly ask some other students to take their place?"

At this the storm-trooper burst out: "The Party does not make deals, it gives orders!"

Professor Fuss flushed red – partly from indignation, partly from embarrassment. He had to let the students go. In the evening Rumer was at the Professor's home. Fuss asked sheepishly:

"How did you like that little scene?"

"Very depressing", said Rumer, "What are you going to do to keep them in their place?"

"There's nothing I can do. Things are beyond that stage already. I am afraid that Germany, the German people, will be split into factions as they were at the time of the wars of religion. If the Nazis get the majority and get into power then we shall have to come to terms with them somehow. Perhaps then everything will get back to normal."

The Fuss family provided an excellent dinner. Fuss played a Schubert piano duet with his wife. It was a quiet, untroubled evening.

"And so, Alisa", said Yuri, going on with his story, "When we parted Fuss and I shook hands and I said to him: 'Thank you, Professor Fuss, for your kind invitation. I would like to wish you all the best. You and I will probably never meet again and if we do our meeting may not be such a friendly affair.' Distracted by my own thoughts, I briefly took his wife's hand, too. Fuss's wife was quite young. I wondered whether I should kiss her hand or not. I decided that, given the circumstances, this was not necessary."

"How stupid you were, my little boy", was Alisa's comment, "They were quite right. Their only mistake was to be so passive. Who will fight for Germany, for the New Order, if German academics hold back? How can one sit on the fence at this most crucial moment in Germany's history?"

"Alisa! What are you saying? Just think what you are saying!"

"What I say is true, my boy. There is nothing I would fear more than to be away from Germany at this time and not to be involved in these great events. I shall share the destiny of my people!"

"Meeting you has given me more pain than joy, after all, although I have loved you so much, Alisa."

Alisa began to stroke Yuri's head. "You are my golden boy ..." And then she calmly mixed the most fearful Nazi slogans with her endearments.

"Alisa, are you not ashamed to repeat these Nazi obscenities? After all, it is no secret that you loved my father and loved us."

"No, no, my golden boy. I am not saying anything against you. You see, there are not only dark-haired Jews, there are also blond Jews. Those are the ones who must be exterminated."

Yuri did not feel like arguing any more with the old woman. There was no sense in appealing to her reason: she was already completely under the spell of Nazism. It is time to get ready to go, he thought, time to go back to Moscow. But he would try one more year in Göttingen. As for Alisa, he would never see her again.

On this trip to Berlin Yuri Borisovich met Einstein for the third time. "... I wanted to see Einstein again and I enquired whether he could receive me", wrote Yuri Borisovich. We met in Einstein's drawing room. He asked me what I was working on. I was at that time fascinated by the mathematical theory of chemical valency and I started to tell Einstein about it. Very soon, however, I noticed that he did not find this subject at all interesting. Feeling embarrassed, I stopped. Einstein was apparently engrossed in his own thoughts. Anyway, he ended the conversation in what seemed to me a rather strange way: "Give my regards to Herr Herglotz." The famous geometrician Herglotz was a professor in Göttingen but I could not make out why Einstein's greetings were to be conveyed to him in particular. This was my last meeting with Einstein. I returned to Moscow a year later." [18, p. 111]

Yuri Borisovich returned to Göttingen in a downcast frame of mind. This whole trip to Berlin had been a disappointment from beginning to end. It had been undertaken because Mila had decided to leave Germany and Yuri had gone to Berlin to see her off. Then there had been the dreadful meeting with Alisa. Finally Einstein, the great Einstein, who had aged beyond belief in the past year, had seemed distracted, with none of the old sparkle in his eyes.

To cap it all, when Yuri Borisovich boarded the Göttingen train in Berlin and settled with a feeling of relief by the window of an empty compartment that he had spotted, the door opened and in came an important gentleman carrying two small suitcases. An eye-catching swastika was pinned to the lapel of his jacket. He slung his cases onto the seat opposite and, throwing up his arm in greeting, introduced himself: "Lawyer Mum – leader of the National Socialist Party in Braunschweig." Rumer's heart stopped – that was all he needed, to share a compartment with a Nazi leader! Nothing would happen, of course – except that they might throw him out of this carriage or abuse him or – well, in fact, all sorts of things might happen.

Suddenly, the lawyer looked at him very closely and with a strikingly kind and polite smile asked:

"Is the gentleman from Argentina?"

"Jawohl! How did you guess so quickly?"

"I can recognize any nationality at a glance. And where do you live in Argentina - on an estate or in a town?"

"On an estate, señor. On the hacienda of my father."

Rumer went on to describe his hacienda in detail, telling the lawyer how many horses he had and about the trouble he had with them. Lawyer Mum soon undressed almost completely and lay down to sleep. As soon as he was safely snoring Rumer collected his few things and changed trains at the very next station.

Good news awaited Rumer in Göttingen. Firstly, the proofs had arrived of the paper on quantum electrodynamics that he had written together with Born. Secondly, his close friend Lev Schnirelmann had arrived in Göttingen from Moscow.

Rumer had been looking forward to his coming. A few months before Gelfond, a mutual friend of theirs from Moscow University, had been in Göttingen. He had produced a brilliant paper on number theory, his particular achievement being to work out a full solution to the Euler-Hilbert problem of transcendental numbers. He had come to Hilbert to show him his solution.

Gelfond had also brought with him a paper written by Lev Schnirelmann in which he had proved Goldbach's famous hypothesis that any even number can be represented in terms of the sums of simple numbers. Rumer and Gelfond had been to Edmund Landau to report on this work. At first Landau did not want to listen. Number theory was his own special field and he had long been trying to prove Goldbach's hypothesis, but without success. All other attempts at proof that he had heard of had also been fruitless. Now here were two young men from Moscow asserting that a third such young man, their comrade, had made the great breakthrough. Looking for and hitting upon a way to get Landau to listen was a long and difficult business – but when he did finally turn his full attention to their account his mind began to work like a machine. "That's it!" he said in amazement, "your fellow-countryman has done it!" On the evening of that same day Edmund Landau addressed the Mathematics Club. The title of his talk was: "Goldbach's hypothesis and Schnirelmann's theorem." Hilbert expressed the desire to invite Schnirelmann himself to Göttingen and immediately sent off a letter to the Soviet embassy.

Now here was Schnirelmann, a 25-year-old professor, short and powerfully built, already balding a little. His whole luggage consisted of his papers, a toothbrush, two changes of underwear and a notebook full of the fairy tales which he used to write. All this fitted easily into his briefcase. At the border he had been asked: "Professor, is your luggage being sent on after you?" "Yes, that's right", said Schnirelmann.

Rumer was delighted to see Lev. The oppressive feeling which had troubled him in Berlin now began to pass. He had to put his own calculations aside for a while so that he could show Lev around. Lev spoke very little German and as he was in great demand he had to be helped. Yuri translated his presentation to the Mathematics Club and went with him to see Courant and Emmi Noether. They went together to the reception arranged in Schnirelmann's honour by Landau.

The young Moscow mathematician conquered Göttingen. The many minor misunderstandings in which he was involved were not held against him. The problem was that for all his exceptional modesty there were two areas in which he could not restrain himself. One of his unfortunate habits was his story-telling. He composed fairy tales in the style of Saltykov-Shchedrin. The subtle play on words and the keen satire in such tales made them comprehensible only to a native Russian – for example, you may remember how in Saltykov-Shchedrin the inhabitants of the town of Glupovo (Imbecilingham) did all kinds of stupid things, such as building a jail using pancakes as mortar. When such tales were told in bad German (Schnirelmann refused to let Yuri translate his beloved stories) they became absolute gibberish.

His second mistake was to enter into desperate arguments with the Germans, even with those who were far from having any Nazi sympathies or even any interest in politics of any kind. He urged them to accept what to him was obvious – that the only possible way forward for mankind was the path of Communism and that the whole German nation had gone mad; some Germans had lost their grip on civilized values, others had become morally blind, and even if there were some who were not wearing swastikas that made no difference! Yuri had to work hard to take the heat out of these confrontations.

On one occasion Schnirelmann's wrath was turned upon a completely harmless reporter from the local newspaper who had been given the task of interviewing the young Russian professor.

"Herr Professor, you are so young and so famous and our newspaper would be honoured if you would give us something we could print."

"Why should I print anything in it when all the time I've been here I have not once read anything in it that made sense? All right, then. Print an advertisement for me – only don't charge me for it."

"Of course, Professor."

"Write this: 'Professor Schnirelmann, now living in Frau Grönau's guest house, magnetizes "Gillette" razor blades.'"

"Yes, Professor, but what for?"

"How should I know? Somebody may need magnetized blades. I'll do it for them. If you don't like that print something else. What about this: 'Professor Schnirelmann from Moscow, who has been stupid enough to come to Germany, volunteers to be sent to a concentration camp.' And which of the camps have you visited, mister journalist?"

Before the flabbergasted reporter could think of a reply Schnirelmann turned to Yuri and said: "What a gloomy outlook!"

About three days before this Gerta Berning had given a party in honour of Yuri's friend. Her sister Mathilda was staying with her at the time. Mathilda never stopped talking all evening. First she babbled on about some important tables that her father, a teacher of gymnastics, had drawn up. He had discovered that the correlation between the age of the jumper and the length of the jump varied from one nationality to another. Suddenly she turned to Rumer:

"And what about you, Rumer? I hope you at least are not afraid of us?"

"Afraid of whom exactly, Mathilda?"

"Oh, you know – the followers of Hitler."

"What? Don't tell me you've joined them!"

"Yes. My husband has joined the Party and I must tell you that the things people say about us are all lies. I ought to know! Well, judge for yourself – there are so many people who are completely baffled by life. They don't know their proper place in the world and all their energy is wasted in searching for work, food and somewhere to live. We help these people – we provide all that they need.

Do you know, there was this Communist in our area. A plumber called Speier. When my husband came to take him off to the camp they sat and calmly smoked a cigarette together. My husband told him he could carry on living at home for a while if he wanted to but Speier refused. So they sat and had a little chat, then they set off without any fuss."

"How have you managed to live here this long?" was all Schirelmann could say when he and Rumer were safely out in the street.

On his last evening they took Schnirelmann to the cinema. They went in a big group. The cinema owner, who was an exceptionally likeable man, was very fond of the young people and knew very well that when they brought a guest he had to show "The Threepenny Opera".

In a way which had surprised him at the time Rumer had come to be on particularly good terms with this cinema owner, whose name was Funke. Back in the days when Mila was still with Rumer the pair of them had once gone to the cinema together. Yuri had caught his jacket on a nail sticking out of the back of his seat. It had ripped out a little flap of cloth. This did not bother him in the least but, to save others from being hurt, he had decided to go and warn the management. At the end of the film he went to the box office and reported that he had been sitting in such-and-such a row in seat number such-and-such and that there was a nail sticking out of it.

"It has not done me any harm but it might cause you trouble."

"Oh, doctor, we know what kind of person you are. You don't just think about yourself. We are so grateful to you!"

The next day the cinema proprietor brought a tailor to Rumer's flat to measure him for a new suit.

"Please let us take your measurements. You suffered damage in our cinema and were kind enough to come and warn us of the problem."

For a long time Rumer would not hear of it but Funke persisted and two weeks later Rumer received a magnificent blue suit which turned out to be just what he needed at the time.

More and more often in recent weeks this same cinema had been closed. When the young people had first seen the sign which announced that that day's programme was cancelled they had gone to the proprietor to find out what was the matter. It turned out that the hall had been hired by the National Socialist Party for one of its meetings. The young people were horrified.

"Herr Funke, what are you doing? You're an intelligent man! Throw them out!"

"That is impossible. I am forced to let them have the hall and to take my leave of you now. You will probably boycott me but there is nothing else I can do."

"No, we shan't boycott you but all the same you should not do such silly things. Why come out openly on the side of people from whom no good whatever can be expected?"

But one man's actions could no longer influence the way the whole town was moving. One day the rumour went round that the Führer himself was coming to Göttingen. Sure enough, a sign soon went up on the doors of Funke's cinema announcing a Nazi meeting which would be addressed by Adolf Hitler.

"How reckless I was at that time. I went off to this meeting, just like that. I was curious about everything then and I did not pause to consider firstly that my life might be in danger and secondly that it might not be worth the risk," said Yuri Borisovich. The idea of going to this meeting was first suggested by Fritz Houtermans. Three of them went along: Rumer, Fritz Houtermans and Hans Helman.

Houtermans was a vivid character. He was a brilliantly gifted physicist and mathematician. As far back as 1929 he had voiced his thoughts about stars as thermo-nuclear energy sources and dreamed of obtaining a thermo-nuclear reaction in laboratory conditions. Blond-haired and blue-eyed, he had a thoroughly "Aryan" face on which innumerable duelling scars told the story of his stormy past as a staunch member of his student society and hard-bitten fighter.

When Rumer once took him along to meet Yakov Ilyich Frenkel, who had come to Göttingen when he was already a scientist of considerable renown, Frenkel was aghast just at the sight of him. Their conversation never got properly started. Houtermans was just not capable of being serious and even when he was discussing scientific matters he looked somehow as if he were juggling in a circus. In the middle of the discussion, he suddenly sprang to his feet and took his leave, excusing himself as he had a rendezvous with a girl (true, he could not remember with which of his passions). When he had departed Frenkel said to Rumer: "Yuri, where did you dig up this Nazi thug? He's got more scars on his face than convolutions in his brain."

Fritz Houtermans' life was to be complicated and confused. His name is linked with the nuclear research later done at Kharkov and with the most dramatic moments in the work on the German atom bomb. He was to be an outsider wherever he went.

Hans Helman, a graduate of the Higher Technical College in Hanover, came to Göttingen to do experimental work in physics. The son of a Prussian regular officer, he had inherited from his father the appearance of the stereotyped Prussian. No less of a duelling enthusiast than Houtermans, he had lost two fingers of his right hand to the sharp rapier of an opponent. He was also a Communist. When the German Communist Party became an illegal organisation and a section of the members declared that they could not accept Party policy and were leaving, Hans Helman was one of those who left. He escaped the attentions of the police and thus was able to join a secret underground organisation and do as much as he had the power to do. He dreamed of getting to Russia and his dream was to come true – in fact his whole life was to be linked with Russia. When he was asked later whether he regretted this he answered: "You would have to be blind not to support Russia now."

And so it was this trio – Fritz Houtermans, Hans Helman and Yuri Rumer – that set off to arrive at their own dear cinema in time for the noon meeting. Youths in brown shirts were standing outside. Hans picked out the one who seemed to him to be in charge, raised his right hand in a fingerless salute and cried: "Heil, Hitler!" Then he took from his pocket an Italian newspaper with a photograph of Il Duce on the front. He explained: "We wish to hear the Führer speak. We have an Italian comrade here with us (pointing to Rumer) who is a representative of the "Corriere della Sera" (the Italian Fascist newspaper) and he absolutely must get into the meeting." Rumer drew himself up and nodded: "Si, compagno."

The lad scrutinized them. His eyes halted on Fritz's picturesque features but could not meet the latter's insolent stare. They were waved towards the box office. There sat a fat man with a butcher's face. He was wearing, naturally, a brown shirt and a swastika armband. He threw up his right arm in greeting and tore off three tickets – 5 Marks each. The young men took their tickets, only to find out that they were not to be admitted to the hall where the Führer would be speaking. They were to go into another hall nearby to which the speech would be relayed over a radio link. Even this second hall was full to bursting. Rumer became very uneasy but Fritz was indignant: "What a swindle, comrades, we want to see the Führer and we shall not put up with this!"

"Don't talk rubbish, Fritz", Rumer said to him, "Let's get out of here while we're in one piece."

"What? And let these swine keep our 15 Marks? Never!" He turned and went back to the ticket office.

"We are poor students, we wanted to see the Führer and we would have paid much more than five Marks to see him – but we were not warned about this. Give us tickets for the main hall. Otherwise you are practically cheating us and the Party is above such conduct."

"Only Party members are allowed in the main hall. Heil Hitler!"

"It is not so vital for Party members to get in. We are just sympathizers as yet – but we have not quite got everything clear in our minds. That's why people like us need to see the Führer!"

Rumer listened with dread as Fritz came out with this drivel. Judging by the way the "butcher's" eyes were becoming bloodshot he was beginning to suspect that they were making fun of him, but no – out came the "Heil Hitler" and the Nazi slogans. Fortunately the bell rang and the cashier grabbed back their tickets, returned their 15 Marks and hurried into the hall.

"What the hell was I doing there?" Rumer cursed himself later, "There'll be no more of that!" But the frequency of Nazi gatherings increased inexorably. They spilled out from the halls into the streets and squares of the little town. It became difficult to avoid these mass meetings.

On one occasion Rumer and Heitler were on their way home from the James Franck Institute, merrily discussing their attempts to repeat Millikan's experiment in Franck's laboratory.

They had arranged to meet Tanya Ehrenfest at the institute that morning. She was the only one who could persuade Franck to allow theoreticians to lay hands on the more delicate instruments. Tanya, the daughter of Paul Ehrenfest, "Tanya with a hyphen" as she was called (Ehrenfest's wife was also Tanya), was the princess of science. In Leningrad she was the guest of Joffe, in Moscow she stayed with Mandelstam and Tamm, in London with Rutherford, in Berlin with Einstein. Tanya was a fine mathematician. Einstein once wrote to her:

"Dear Tanya, I've got mixed up in all these indices. Come to Berlin as soon as you can and sort them out."

Franck's laboratory was magnificently equipped. Everything was precisely calibrated and adjusted. This made it possible for anyone to come along and repeat complex experiments. The two theoreticians and their friend the mathematician set about repeating Millikan's experiment. They first decided which of them would be responsible for what: Tanya would look after the instruments, Rumer would measure the intervals and Heitler would record the frequency of the intervals. Rumer found it useful to count out "Gop-gop-gop" before pressing the button on the stop-clock.

"Will you stop your idiotic gop-gop!" Tanya said to him, "It's driving me mad. Count some other way. Knock on the table if you must."

Rumer started knocking instead – and everything rattled with each blow. In came Franck to ask how things were going.

"We've decided to bang on the table instead of saying 'gop'."

Franck looked at the quivering needles of his exquisitely delicate instruments and was horrified:

"I would rather have ten complete morons in my laboratory than two theoreticians!" he cried – and threw them out.

On the way back from the Institute Rumer and Heitler were in a jovial mood. They walked along in their normal fashion: Rumer on the road, Heitler on the pavement, so as to compensate for the immense difference in their heights. Suddenly they noticed that a cluster of people had formed in the square and that everyone around them was running in that direction. They also ran and were rewarded with the sight of Goering himself roaring a rabble-rousing speech at the crowd. He was fat, with a mane like a lion and covered with medals. At first his listeners behaved normally. Only a few raised their hands in salute along with Goering. Then the temperature began to rise. Goering's voice rose to a scream:

"I have no Fatherland! We have no Fatherland! The Fatherland has been betrayed! They pinned these shining baubles on me – but what use are they? I don't need them. I need my Fatherland!" And with a trembling hand he grabbed at his chest, ripped off the medals and threw them to the ground.

"The most terrifying thing was that his speech was so convincing," recalled Yuri Borisovich, "that the crowd began to believe in what he said and to share his hot anger. And every time he called out "Heil!" more and more arms shot up."

The young men scarcely managed to extricate themselves from the crowd and crawl home, completely shattered. It was clear now that they had to leave Germany – and the sooner the better.

"I shall leave neither tomorrow nor the next day, but in a week from now", thought Yuri Borisovich. A week later he did indeed leave.

The evening before his departure there was a party in Frau Grönau's lodging house. It was a sad party. Such parties had been held more and more often recently. One by one they were all leaving for different destinations. All Rumer's friends came to see him off at the station. Naturally, the Borns were there, too.

Just in front of the station, when Yuri Borisovich got out of the car, he saw a little boy. The boy noticed the long, tall man climbing awkwardly out of the car and moved closer. Rumer nodded to him and smiled and the chubby little lad threw up his arm in a proud Nazi salute. He was too young to know any other greeting.

The heart of a reactor (uranium cubes)

Chapter 11

The Choice

In January 1933 Adolf Hitler was sworn in as Chancellor of Germany by President von Hindenburg. In March the new Minister of Education, Hitler's appointee Rust, issued an order dismissing all Jewish academics from all the educational establishments and research institutes in Germany. In April a list of the people considered undesirable by the new regime appeared in the newspapers. Einstein, at that time in Belgium, announced his resignation from the Berlin Academy.

In his speeches on the radio Goebbels held the staff and students responsible for Einstein's behaviour. In response to the "tactlessness of this man of whom all Germany had been so fond and who now works against us abroad" [37, p. 258] Rust issued another order banning any mention of Einstein's name in the educational system. The threats directed at Einstein went as far as an open demand for physical violence against him and those like him.

The German universities were cut down as if by a scythe. Old Hilbert was left almost completely alone in Göttingen. He died in 1944 and no more than ten people followed his coffin.

Tens, then hundreds of people fled from Germany, among them Courant, Emmi Noether, Max Born, James Franck. Some were forced to leave while others, though not themselves threatened, left in protest against the policy of the new regime.

1931 is considered to be the year when American physics was born. This was the year when the AIP, the American Institute of Physics, was founded. This body received considerable financial support and was able to bring under one umbrella the research being done by many small, scattered groups of scientists. The Institute's other aim was to establish links between the major industrial firms and coordinate their research.

The "50th anniversary of American physics" was celebrated in 1981. This was the frank title chosen for the November issue of "Physics Today" [32]. On the cover were photographs of the Americans who won Nobel Prizes for physics between 1931 and 1981. There were no fewer than 40 photographs – proof of an impressive achievement. And in large measure, as the journal acknowledges, this achievement was made possible by that "massive flow of intellectual capital which crossed the Atlantic from Europe".

In 1931, on the occasion of the founding of the AIP, Millikan (Nobel Prizewinner 1924) predicted that from then on, "the United States and Germany will be the world leaders in science." But he was wrong. Less than two years later the majority of German scientists had "arrived on the shores of America" and Germany was out of the running.

Of course, there were scientists who stayed in Germany. Max Planck and Max von Laue remained in Berlin. Max Planck spoke out fearlessly in defence of the exiles and stood up to the new order. When he went to see Hitler their conversation ended with the latter in hysterics. Planck's friends quite rightly feared for his life.

Otto Hahn, Lise Meitner, Jordan and Weizsäcker all stayed in Germany. So did Heisenberg, whose activities over two decades were riddled with contradictions. His name is inevitably linked with the decline of German science and with the attempt to build the Nazi atom bomb.

If all the photographs taken of Heisenberg in different phases of his complex life are laid out in order, dramatic changes in his facial expression, his eyes and his posture become apparent. At 17 he is a storm-trooper in khaki shorts, a member of one of the volunteer detachments which fought the workers' and soldiers' revolution of 1919.

Between the "Freikorps" trooper and the creator of quantum mechanics little difference can be detected; there is still almost the same proud bearing and the same cheerful, straightforward look in his eyes.

Over the next ten years, in fact, there is little change. In the postwar photographs, however, Heisenberg is suddenly old and bent, a completely different person. In this series of pictures there are none from the war years, when Heisenberg was in charge of the work on the atomic bomb.

Judging by Heisenberg's own memoirs, he was at first torn between staying in Germany and leaving. There is an account of this period of vacillation and the final decision to stay in his book "The Part and the Whole" [39]. This book, though in essence the story of Heisenberg's own life and work, reads like a chronicle of the rise of modern physics. It is constructed in the form of imagined dialogues with Niels Bohr, Pauli, Planck and others – the choice of interlocutor depending on which scientific problem is under discussion.

Heisenberg chose as his introductory epigraph the words of the ancient Greek historian Thucydides: "As for the conversations which were held then and in which I participated directly myself, I found that I could not retain in my memory the exact import of everything that was said. I therefore put into the mouths of individuals the words which, as I understand it, they could have said in the given circumstances. As far as I possibly could I tried to follow the actual train of thought of the speakers." [39, p. 9]

In this book one chapter is entitled "The Revolution and the University 1933". [39, p. 187]. The very title of this chapter, in which Heisenberg calls the Nazi putsch a Revolution, puts the reader on guard. The first and larger part of the chapter is an imagined conversation between Heisenberg and a Nazi student; the second part is a chat with Max Planck.

The first part begins with the student's question: "Why is it that you do not want to have anything to do with us?" In the long drawn-out conversation that follows the student tries to convince Heisenberg that Nazi ideology is based on truth and that the way forward chosen by the Nazis is the only possible one. Heisenberg objects – but all his objections shatter against the student's iron logic. In places it is indeed hard to argue with him. "The fact that many older and less vigorous people lost everything and were left to starve did not cause any outcry. The government had plenty of money. The rich got richer and the poor got poorer ... Was the government concerned about the poor?" asked the student. "You must admit that we are doing better. We talk to and listen to the workers and train with them in the same storm-trooper units. We collect food and woollen clothes for the poor and march with the workers in joint demonstrations. We feel that they are glad that we are involved in their lives. This is a real improvement ..."

Heisenberg's reaction is a simple one: "And do you believe that your Führer, Adolf Hitler, is an honest man?"

"I can imagine that you are not drawn to Hitler, that to you he seems too primitive" – and then the confident prediction that, despite everything, Hitler would win greater recognition, even among Germany's enemies, than any of his predecessors.

"We do not fully understand each other on this point", Heisenberg tries to reassure him, going on to say that every individual should try to live a peaceful and creative life, that a political movement must not be justified and evaluated by its stated ends but by its chosen means, that in this case the chosen means are bad and will bring only woe.

"But nevertheless you must admit that gentle methods will get us nowhere ... You criticize us for following a man whose methods you do not approve of. I also see his antisemitism as the negative aspect of our movement and I hope he will soon leave it behind him. But is there one figure from the old world, one of the old professors who now complain about the Revolution, who has tried to show us, the younger generation, a better way which would take us by better means to the right end?"

"We, the older generation, cannot give any advice, for the simple reason that we have none to give – except the unexciting assertion that one must do one's work conscientiously and decently in the hope that a good example will, in the end, have a good effect."

"All you want is the past, the things of yesterday, and you condemn any attempt to change them ... So what right have you to stand up for revolutionary ideas in science? Remember that the theory of relativity and quantum theory were a radical break with everything that went before."

Heisenberg explains to the student the real character of the revolution in science, describing Planck's struggle with himself over many years and his attempts to repair the breach in physics which he himself had made. It is difficult to compare scientific and social revolutions, says Heisenberg, whose main point is that a revolution can be successful only if it restricts itself to the solution of certain concrete problems, since the scope for real change is limited: "You will remember the revolution proclaimed 2,000 years ago by Jesus Christ who said: "I have come not to destroy the Law but to bring it to fulfilment."

The conversation was now losing its fervour and tapering away to a natural close. "My visitor was already getting ready to leave when I asked him if he would like me to play the last part of Schumann's concerto as far as is possible without the orchestra. He was glad to accept the offer and when he finally left I had the impression that his attitude towards me was friendly.

In the week which followed this conversation the attacks on the university became more determined and intimidating ... We considered resigning from our posts and convincing as many as possible of our colleagues that they should do the same. But before taking this step I wanted to talk the whole matter through with some of our older colleagues that we knew we could trust. I requested a meeting with Max Planck ...

Planck seemed to have aged by many years since the last time I had seen him. Deep wrinkles now lined his thin face. There was suffering in his smile of greeting and he looked exhausted.

'So you have come to me for advice on political matters?' he began our conversation, 'Well, I'm afraid I can't give advice any more. I have lost all hope that the disaster descending on Germany and the German universities can be halted. Before you tell me about the violence in Leipzig, which was probably just as bad as that here in Berlin, I want to tell you about the talk I had with Hitler a few days ago. I had hoped to make clear to him what damage is being done to the universities of Germany, and to the physical sciences in particular, by the persecution of our Jewish colleagues. I wanted to make him see how senseless this policy is ... But Hitler did not show any sign of understanding. Worse than that, there is simply no language in which one can communicate with this man. It seemed to me that Hitler has lost all contact with the real world. All that is said to him he takes at best as a tiresome hindrance which he immediately seeks to remove by declaiming over and over again the same phrases about the disintegration of intellectual life over the past 14 years, about the vital need to halt this slide before it is too late, etc, etc. Moreover, I received the dispiriting impression that he himself believes all this nonsense ... and since he is obsessed with these so-called ideas of his, he is not open to any rational objections. I am convinced that he will lead Germany to a dreadful catastrophe.'"

Heisenberg then reported on the events in Leipzig and told Planck that several members of the faculty were planning to resign in protest.

'I am glad,' said Planck, 'that you are still optimistic and believe that such actions might yet avert disaster. However, I am afraid that you over-estimate the influence of universities and of intellectuals. Society as a whole will hear practically nothing about your protest. The newspapers will either ignore your resignation completely or will report it in such a hostile manner that no one will even think of letting it change their thinking. Don't you see, once the avalanche has started there is no stopping it. If you resign then at best you will have to look around for a post abroad. As for what could happen in the worst case, I had better not mention that. Abroad you would simply be yet another member of that multitude who emigrate and start job-hunting. And maybe you will, in an indirect fashion, take the job of another who needs it more than you do ... After the catastrophe you will be able, if you so wish, to come back to Germany – and with a clean conscience, knowing that you did not enter into any compromises with those who devastated Germany ...

However, if you stay here and do not resign, then the tasks which face you will be of another kind altogether; you will not be able to prevent disaster and, in order to survive, you will have to compromise all the time ... and those compromises which you will be forced to make will afterwards be judged and perhaps you will be punished for them ... But in this awful situation, which we in Germany are now in, it is already impossible to act correctly. Whatever your decision, you will in any case be involved in wrong-doing.

Anyway, when you make your decision, think about the time which will come afterwards.'

This, in a few words, was the gist of Planck's response, but we must remember that Heisenberg quotes only fragments of what Planck said and that he followed the principle of Thucydides when reporting the words of others.

In the train on the way back to Leipzig Heisenberg pondered on his conversation with Planck: "I was tormented by this question of whether I should emigrate or stay. I was almost envious of those who were forced to leave ... at least they did not have to make this terrible choice ... What is the right thing to do? To leave one's homeland and flee from the infection or to stay and try to help one's sick country, even if no hope remains. One had to think about the time after the catastrophe. Planck had said this, and it was this that persuaded me that I should try to form pockets of resistance, gather the young people together and if possible bring them alive through the catastrophe. Then to rebuild everything again when the crisis was over. This was the task which Max Planck had had in mind. It would inevitably mean making compromises which could later lead to punishment under the law, or worse. But at least this was a clearly formulated task. By the time I got back to Leipzig my decision was made – i.e. to remain, at least for the moment, in Germany and see where this path would lead me."

So, having made his own personal choice – to stay and see where "this path" would lead – Heisenberg was in fact looking ahead to what Max Born called "the present horrifyingly strange state of affairs in which humanity ... is faced with the stark choice between peace and self-annihilation". "This path" would lead Heisenberg to the post of supervisor of Hitler's uranium project.

It is hardly necessary to speculate as to what the atomic bomb would have meant in the hands of Adolf Hitler. It is enough to note that just this possibility gave the spur to the physicists who had been deported from Germany and who knew only too well how feasible the atom bomb was, especially with Heisenberg in charge, and urged them on to frantic efforts in the race to create the American bomb. No one regarded the Soviet Union as a serious contender in this area of research at that time.

At the very end of 1938 Hahn and Strassmann discovered the effect which Lise Meitner called the fission of uranium. The article in which Hahn and Strassmann described this effect was published on 6 January 1939 in the journal "Naturwissenschaft". Physicists all over the world understood its significance immediately: if a uranium nucleus splits into two light nuclei, and this process can be maintained, then according to Einstein's theory the colossal energy locked up in the nucleus would be released.

A conference in Berlin in 1942, which was devoted to the tasks faced by the research council of the Reich, was attended by, among others, the supervisors of the German uranium project. Goering, in a characteristic address, shouted reproaches at the physicists: "What is the cost of the impatience and haste with which this or that scientist trumpets his discovery to the whole world? They simply cannot restrain themselves and are as anxious as a man with an over-full bladder not to delay for a moment. Magnificent! Wonderful! Everybody gets to hear about it!" [40, p. 153]

But by this time it was already too late. In that same year, 1942, the world's first reactor went into operation under the grandstands of Chicago's university stadium. Hahn and Strassmann, when they published their work in such "haste", were not thinking about politics and, moreover, could not have predicted what their discovery would mean for the human race. Of course, Hahn understood perfectly well what the implications of the discovery were but neither he nor anyone else could have supposed that a mere six years would elapse between the complex laboratory experiments, the precise proof of the bare fact that this physical effect existed, and the bombing of Hiroshima.

And so, 6 January 1939 saw the appearance of the journal containing the article by Hahn and Strassmann. After the Christmas holidays Niels Bohr got ready for a trip to the United States. On 26 January he was already reporting the results of Hahn's experiment to the 5th Washington Conference on Theoretical Physics. Bohr's presentation made such a profound impression that some of the physicists, still in their evening dress, dashed back to their laboratories as soon as he stopped speaking to check his assertions. 24 hours later the American newspapers were already reporting on these experiments.

The day before Bohr's report to the conference Hahn and Strassmann, with that same "haste", had sent off a new article to "Naturwissenschaft". This was "Proof of the Formation of Active Isotopes of Barium by Neutron Bombardment of Uranium and Thorium" with the subtitle "Further Active Products of the Fission of Uranium". It was this subtitle which reflected the article's main content – this being the revelation that, when a uranium nucleus is split into two almost equal parts, neutrons are released along with a massive amount of energy. The conclusion was self-evident: these liberated neutrons can themselves participate in the fission process and cause the splitting of adjoining uranium nuclei. If the number of released neutrons should be greater than one then the possibility of an avalanche effect – i.e. a chain reaction – seemed obvious.

In April Bohr and Wheeler, and independently of them Yakov Ilyich Frenkel, formulated the theory of the splitting of nuclei by slow neutrons and predicted spontaneous (self-perpetuating) fission of nuclei. Soon the possibility of the spread of a nuclear chain reaction was firmly established (by Szilard and Fermi in America, Joliot-Curie in France, Zeldovich, Khariton, Leipunsky in the Soviet Union and others). The energy spectrum of fission neutrons was measured and delayed neutrons were discovered experimentally.

It became apparent that in the fission of the uranium nucleus the fission products rid themselves of excess neutrons in various ways. Most of the neutrons are released at the moment of fission – these are the so-called "prompt" neutrons – but there are also neutrons which are released a little later – these are the "delayed" neutrons. So, if only prompt neutrons were released it would be impossible to control a chain reaction, while if there were only delayed neutrons then it would be impossible to build an atomic bomb, although the control of a chain reaction would be much simpler.

The need now was to calculate precisely the proportions of the different neutrons and then, depending on the result and on the aim of the exercise, find a way either to control the chain reaction or to produce an explosion of unheard-of violence. The above-mentioned elementary theory of nuclear fission, though still imperfect and rather crude, provided a good description of the fundamental nature of the fission reaction and, very importantly, predicted the different character of the course of reactions in two different isotopes of uranium – uranium 238 and uranium 235.

The first of these is the most common uranium isotope. It constitutes 99.3% of natural uranium whereas only 0.7% of natural uranium is uranium 235. It is precisely the different fission characteristics of these two isotopes which present the problems and offer the solutions as regards both the controllable chain reaction and the explosion.

The uranium 238 nucleus splits into two fragments only when hit by neutrons which possess an energy which exceeds a particular threshold level. Neutrons with lower energies are readily absorbed by these nuclei, with the result that the uranium nuclei are transformed into nuclei of transuranic elements.

As for uranium 235, here the theory predicted that uranium 235 nuclei would split when hit by neutrons possessing any level of energy. Moreover, it became clear that slow neutrons were 250 times more likely to hit uranium 235 nuclei than they were to be captured by uranium 238 nuclei. In other words, in order to set off a self-sustaining chain reaction it was necessary firstly to slow down the fission neutrons and secondly to increase the proportion of uranium 235.

The second task, that of increasing the percentage of uranium 235, turned out to be extraordinarily difficult. The selection and insertion of the moderating material which would slow down the neutrons was not easy, either. Now, with nuclear power stations and research reactors of widely differing types and sizes in operation and the nuclear bomb in existence, everything seems simple and it is easy to explain the principles on which they work. However, the effort to produce the first chain reaction and the first atom bomb demanded the whole arsenal of knowledge accumulated up to that time and the mobilization of vast scientific, technical and financial resources. Nevertheless, the breakthrough was made.

The first controlled chain reaction was achieved on 2 December 1942 under the supervision of Enrico Fermi, using a uranium-graphite pile. The pile had the shape of an ellipsoid with a polar radius of 309 cm and an equatorial radius of 388 cm.

"When we discussed the question of choosing the right moderator for our pile," wrote Enrico Fermi, "we came to the conclusion that graphite was the most promising material, mainly because it was readily available." Simple enough, it would seem. As we shall see later the choice of moderator was to be one of the most insuperable obstacles in the way of the German uranium project. He goes on: "The erection of the pile took a little over a month ... Long before the structure attained the originally envisaged dimensions the neutron density readings showed that critical mass would soon be reached."

In practical terms the build-up to critical size meant the addition of layer upon layer of uranium and graphite. Plates with strips of cadmium attached to them were inserted into the structure. Cadmium is a very effective absorber of neutrons. As the pile was erected these cadmium strips were padlocked in place and unlocked when the cadmium needed to be removed for the next experiment. On 2 December 1942 the reaction was achieved. "In order to get this new technology to the point where it could be used industrially", wrote Fermi, "a large number of scientific and technical developments were needed ... The first industrial piles were in operation roughly two years after the first experimental pile was started up at Henford."

These industrial piles were put into operation with one purpose only: to create plutonium – the basic material needed to make the bomb. To this end several mighty installations were at work, using three different methods of separating the isotopes of uranium. The uranium 235 which the plants had produced by the middle of 1945, and which had swallowed up millions of dollars, was sufficient for only one bomb – for the one which decided the fate of Hiroshima. The other two bombs used plutonium.

At Berkeley in 1940 McMillan and Abelson bombarded uranium with neutrons to produce the first transuranic element – neptunium-239. The half-life of neptunium turned out to be 2.33 days. In 1941 McMillan and Seaborg synthesized a second transuranic element – plutonium-239, Fermi's ekaosmium. These transuranic elements, which come immediately after uranium in Mendeleev's table, were named after the planets Neptune and Pluto on the analogy of the solar system.

McMillan and Seaborg and their co-workers also proved that the plutonium-239 nucleus is split by slow neutrons in the same way as uranium-235 and is likewise subject to spontaneous fission. If the half-life of plutonium were found to be long enough to make it possible to collect the small amount which constitutes the "critical mass" then plutonium, instead of a rare isotope of uranium, could be used for the bomb. The half-life of plutonium turned out to be 24,400 years. Plutonium, moreover, accumulates in a reactor which is in use over a long period and, leaving aside the complex technology of the uranium-graphite reactor which produces the plutonium, the plutonium bomb itself is such a simple device that all the scientists involved in nuclear physics saw it as an obvious possibility.

On 24 December 1946 a chain reaction was initiated in Kurchatov's laboratory and on 29 August 1949 the first Soviet atom bomb was tested: a uranium-graphite reactor, a plutonium bomb. Let us remember that in 1939 Frenkel had developed the theory of nuclear fission in uranium at the same time as Bohr and Wheeler. In the same year Zeldovich and Khariton formulated the theory of the nuclear chain reaction. In 1940 Flerov and Petrzak discovered the phenomenon of spontaneous fission of uranium-235 nuclei.

At the beginning of 1943 Kurchatov presented a review of "The Uranium Problem" which summarized the main results of the work done up to that point, from that on the structure of the atom and the first experiments in the artificial transformation of elements up to the theory of the nuclear chain reaction. The concluding sections of the review covered the following points: the splitting by neutrons of uranium-235 nuclei, the splitting by neutrons of uranium-238 nuclei, the formation of "ekarhenium" and "ekaosmium" (neptunium and plutonium), concrete proposals for the exploitation of the disintegration of uranium. This same review gives details of the theoretical calculations of Soviet physicists up to 1941 (!) relating to chain reactions in ordinary metallic uranium, in metallic uranium-235, in a mixture of ordinary uranium and heavy water and in a mixture of ordinary uranium and carbon (!!).

The facts are well known. They are cited here only as a reminder that for the professionals there was nothing secret about the atomic bomb. There were difficulties, fantastic difficulties, in its technical realization. These difficulties were overcome mainly thanks to one absolutely obvious fact: the scientists of Hitler's Germany were far ahead in this field and there was no time to lose. Otto Hahn and Werner Heisenberg had stayed in Germany. Confidence in the superiority of German science was still unshaken. German technology was excellent, Germany's chemical industry was the mightiest in the world – and Germany was ruled by a fanatic.

In 1947 Goudsmit wrote: "It is now more than two years since the end of the war but even now not only people outside science but also many of our scientists and military experts think that it was only by a miracle, with our fate hanging by a single thread, that in the desperate race against the Germans to master the secret of the atom bomb we managed to get there first." [41, p. 7]

In fact there was no miracle. It is absolutely true, of course, that the Germans were the first to begin research in this field, that they were the first (in April 1939!) to inform their government of the "brilliant prospects" for a revolutionary weapon, the power of which would be "quite sufficient to devastate and blast into the atmosphere even the greatest city", that only Germany – and this was even before the war – had a military establishment totally dedicated to the "uranium project", that their "Preliminary Working Programme of Initial Experiments relating to the Exploitation of Nuclear Fission" is dated 20 September 1939.

Fortunately, however, for the whole of mankind, from that very April of 1939 right to the end of the war a feverish game of leap-frog was played in Germany. The project was riddled with intrigue and enmity – each of the many hopefuls wanted to be in control and, if not in control, then at least to carve off a piece of the "atomic pie" and then a bit more.

One cannot help being reminded here of the ancient parable, which has been handed down in collections of wise tales, about the old tsar who called his sons to his death-bed and gave each of them (in the Georgian version) a tough, springy vine-rod. "Break it", ordered the tsar. The sons broke their rods. "Now break this", said the tsar and handed them a bundle of vine-rods. It is a blessing that the zealous participants in the German uranium project never managed to unite in this way before the war was lost.

And yet each of them, including Heisenberg, did his very best. What is most amazing is that, even under Heisenberg's leadership, the German scientists never hit on graphite as a neutron moderator. They rejected it at the very start – and this was their first fatal mistake. Their second lay in the fact that they did not arrive at the plutonium solution (actually, the full story is much more complex – we shall return to it later) and spent all their time striving to create a uranium-235 bomb. Moreover, they envisaged that the bomb would be a kind of reactor compact enough to be loaded into an aircraft.

It would be wrong, of course, to think that only graphite can be used as a moderating material or that bombs can only be made from plutonium. There is no doubt that heavy water is an excellent moderating agent and deuterium is widely used in modern reactors. But at that time, in conditions of frantic urgency, the production of deuterium in large quantities was an even more difficult business than the production of really pure graphite – or rather, it was much more expensive.

As for the uranium-235 isotope, its separation from natural uranium was a task of simply unimaginable complexity. The problem is that there are no chemical means whereby different isotopes of the same element can be separated and so methods based on physical effects had to be devised. This was a truly Herculean task. Enormous efforts resulted in the invention of various methods of separating the isotopes but they all proved to be remarkably ineffective and to involve huge technical difficulties. And all the time, while in America great industrial plants were built to serve a single purpose and a single laboratory, the same game of leap-frog went on in Germany. It began in 1939 when almost simultaneously two ministries, the Ministry of War and the Ministry of Education, each decided to create its own uranium research programme. In the case of the Ministry of Education the main instigator was Abraham Esau, who had with great difficulty convinced the minister of the importance of uranium research. But Esau knew very well that the support of the Minister of Education was not enough and he also approached the Ministry of War. By that time the war ministry had already received the famous letter from the Hamburg physicists Harteck and Grot concerning the possibility of preparing an explosive substance of hitherto unimagined power. The letter ended with a simple conclusion: "The first country which is able to take possession on the practical level of the achievements of modern physics will acquire absolute supremacy over all others." [40, p. 44]

The ministry did not reply immediately; the letter was passed around from hand to hand. Its first recipient was the chief of the Army's armaments department, General Becker. From him it went to the head of the research department, Professor Schumann (who was military advisor to Keitel) and in the end it was passed on to the Army's explosives expert, Kurt Diebner.

Diebner immediately realized the importance of the letter but despite his persistent and eloquent demands, he got no response from the higher authorities. It was only when Esau appeared on the scene in his attempt to interest two ministries at once in uranium research that a research bureau was set up with Diebner as its head.

Heisenberg, who was working hard on theoretical calculations pertaining to nuclear chain reactions and at the same time making plans for his own experimental installation at Leipzig, became Diebner's scientific consultant. But to the very end, until the Spring of 1945, there was unceasing hostility and rivalry between Diebner's people and Heisenberg's group of academics. More than once letters, all ending, of course, with the obligatory "Heil Hitler!", were sent to the higher authorities complaining and protesting about the endless difficulties that arose in connection with the allocation of materials.

There were intrigues on every side. Esau, though spurned in the rudest possible manner by the war ministry, did not give up. The heights he had reached did not satisfy him at all and by the end of 1942, having enlisted the support of Goering, he took up the post of head of the German uranium project. However, by the beginning of 1944, thanks to the secret intervention of the armaments minister Speer, Esau had been removed. His successor was Walter Gerlach.

While this fuss in high places was going on, ever renewed by the crossfire between the rival groups that clustered around Diebner, Esau and Heisenberg, yet another contender in the race to solve the uranium problem came to join the "common" cause. This was the brilliant German engineer Baron von Ardenne.

He had been forbidden to seek the necessary funds in the ministries over which Goering presided and Rust, the minister of education, wanted to hear no more of any "uranium project". The Baron next turned to the large research department attached to the communications ministry.

Ardenne managed to arrange a meeting with Ohnesorge, the communications minister to whom he explained the importance of nuclear research, adding that uranium reactors had massive future potential as "a source of power suitable for use in ships." The minister was shaken. He immediately obtained an audience with Hitler and conveyed to him the gist of what he had heard. It was 1940. Most of Europe was under German occupation and plans for the blitzkrieg attack on Russia had been finalized. It was a time of triumph for Hitler. The Führer humiliated Ohnesorge in front of those present, saying that while his ministers and generals were "racking their brains to find a way of winning the war, the head of the Post Office has brought a ready-made solution." [41, p. 130]

Stung by Hitler's mockery, Ohnesorge immediately incorporated Baron von Ardenne's project in his own ministry's research programme.

The formation of this new group was greeted with hostility by the two teams which already existed. In the autumn of 1940 Weizsäcker walked into Ardenne's laboratory and exhorted him to give up his work, saying that in Heisenberg's opinion the atomic bomb could not be built. The only thing that united the groups and the warning factions within the groups was the unquestioned authority of Heisenberg.

As for von Ardenne's team, Heisenberg's scornful attitude toward everything that they did was to prove fatal for the whole German uranium project. It was the Baron's team which came up with the idea that uranium-235 was not the only material with which an atomic bomb might be made.

Fritz Houtermans came to Ardenne's laboratory at the beginning of 1941. After working there for eight months Houtermans presented a paper in which he offered a remarkably clear and rigourous answer to all the main questions involved in the "atomic problem". He completed the calculations relating to a chain reaction brought about by fast neutrons, worked out the critical mass of uranium-235 and, most important of all, he considered the reactor as "a device for the transformation of elements" and gave special emphasis to the fact that when a reactor is operating a new radioactive substance is formed. This new element has similar properties to those of uranium-235 and is also suitable for bomb-making.

In this way Houtermans, who did not suspect that the new substance had already been discovered beyond the ocean and dubbed plutonium, had hit upon the solution which was the best available at the time. Houtermans also made a parallel study of all the various ways of separating isotopes. Analysing them from an economic point of view, he showed that the collection of uranium-235 was an almost hopeless task, whereas the new alternative – the collection of the new radioactive substance – both solved the problem and removed the need to separate isotopes.

It is hard to imagine a more contradictory nature than that of Houtermans. An opponent of Nazism who had been in various prisons, he first fled from Germany to England and from there to the Soviet Union. He found his way to Kharkov and there threw himself into his work. For some reason which is not known he was then deported from the USSR. In Germany he was an outlaw. He was constantly watched by the Gestapo and banned from working in any state institution. Baron von Ardenne risked his own head by taking Houtermans on as a co-worker (Ardenne's laboratory was considered to be a private establishment). As we have seen, Houtermans did not take long to produce results.

Knowing full well what a threat they posed to the whole human race, he did not attempt to keep his conclusions secret. On the contrary, he informed both Heisenberg and Diebner. Later, with Diebner and Schumann, he was to turn up in Kharkov in the uniform of an S.S. officer and try to persuade his old Kharkov colleagues to cooperate with the Germans. Moreover, Houtermans twice (!) published his paper in secret reports – that is, he tried with all his might to bring it to concrete realization. Even that is not all. In 1943, while on a visit to Switzerland, he somehow managed to send a telegram to Chicago – and it got through. The message consisted of the words: "Hurry, we are on the right track!" [42, p. 102]

Interestingly, this telegram frightened the American scientists not so much by its content (they were already convinced, anyway, that the Germans were far ahead and, naturally, on the right track) as by its demonstration that the Germans' knew about the "secret research" going on in Chicago.

In actual fact, it was just in his assertion that the Germans were "on the right track", that Houtermans was most grievously mistaken. Neither Heisenberg nor anyone else paid any attention to Houtermans' results. All the uncoordinated efforts of the German researchers, of the physicists and chemists working in both industry and in academic laboratories, were thrown into the separation of isotopes and the production of heavy water.

Since its beginning in 1939 the research into chain reactions had proceeded at full speed. Heisenberg had worked in Leipzig and then in Berlin. Bothe's team had worked in Heidelberg and Harteck's in Hamburg. There is no need to list all the teams – only to note that they had all worked flat out.

In Norway in 1940 the Germans had seized the world's only heavy water plant. After the defeat of France the magnificent Joliot-Curie laboratory with its brand-new American cyclotron and several thousand tons of uranium fell into the hands of the German occupiers. In occupied Czechoslovakia the German firm "Auer Gesellschaft" lost no time in exploiting the uranium deposits. By the end of 1940 a production line was established, with one tonne of pure uranium being turned out each month. The Germans took two tonnes of sodium uranate from Belgium.

In Berlin in December 1940 Heisenberg, Weizsäcker and Wirtz began assembling the first atomic pile. The "Degussa" factory in Frankfurt received an order from Heisenberg for uranium plates one centimetre thick. Heisenberg had settled on a construction consisting of uranium plates with heavy water as the moderator.

The same firm received an order from Diebner for cubic blocks of uranium. The promising result achieved by Diebner came as a shock to Heisenberg: towards the end of 1942 Diebner's reactor achieved the highest coefficient of neutron propagation yet reached in Germany. In his report on these results Diebner wrote: "The enlargement of the reactor will definitely bring it to a critical state. All that remains to be clarified is the exact amount of uranium and heavy water required." [40, p. 199] It was the required amount of uranium and heavy water which each of the groups now tried to determine for itself.

At the beginning of March 1943 a coded message was received in London: "High concentration installation in Vemork completely destroyed on night of 27/28 February. Gunnerside." Operation Gunnerside, the aim of which was to blow up the unique heavy

water plant, was planned in England and carried out by four Norwegian patriots. In the history of espionage and clandestine operations it is not regarded as a major event. Perhaps that judgement is
correct.

Let us note, however, that this was the second attempt. The first
operation, also prepared in England, ended tragically. It cost more
than thirty lives and the natural reaction on the part of the Germans
was to strengthen the plant's defences. This they did with real German
thoroughness, using everything from anti-aircraft guns to minefields.
Nevertheless the plant was blown up.

Just how important the uranium project was in the estimation
of the German high command can be judged by the fact that the
Vemork plant was back on stream within two months, instead of the
expected two years. In the autumn of 1943 American "Liberator"
bombers operating from England dropped more than 700 bombs on
Vemork. The smoke-screen generators placed around the plant by
the Germans had their effect and the heavy raid did not achieve the
desired result: the most vital targets remained undamaged and the
heavy water plant itself emerged unscathed.

However, the main achievement of the raid was the destruction
of the power station and this was enough to paralyse the plant. The
Germans decided to dismantle it and transport it to Germany. They
also planned to take what remained of the heavy water that had been
produced. This undertaking required extreme caution. Himmler
personally gave orders that a special fighter squadron should provide
air defence for Vemork and the removal operation was carefully
disguised. Nevertheless, it was wrecked by the Norwegian Resistance
who sank the ferry carrying the last 39 barrels of heavy water in the
deepest part of Lake Tinn.

The shortage of heavy water was keenly felt by the German atom
scientists and they did their best to make good this deficiency. They
tried to modify the design of the pile, to increase the proportion of
uranium-235 and to organise local production of heavy water.

Towards the end of 1944 Wirtz's team in Berlin-Dalem made the first use of graphite, but only as a neutron reflector. By that time this was the only team left in Berlin. In the summer of 1944 American and British aircraft had bombed Munich and Dresden and the RAF kept up constant raids on Berlin. The success of the Red Army's latest offensive meant that the fall of Berlin could not be long delayed.

During the summer Diebner's team was evacuated to the village of Stadthilm. The other part of the Berlin-Dalem institute was moved to the village of Heigerloch near Heisingen. Heisenberg and Wirtz remained for the moment in Berlin where they were completing assembly of the largest reactor. On 29 January 1945 everything was ready for the decisive experiment – but it proved impossible to go ahead with it. On 30 January Gerlach ordered that the reactor be dismantled. The drone of bombers could be heard all over Germany and ground fighting was going on in all but a few parts of German territory. It was in these conditions that a new atomic research centre was formed in the little village of Heigerloch.

Here on the last day of February 1945, under the supervision of Heisenberg and Wirtz, the reactor was started up. The instrument readings were so promising that, unable to conceal his excitement, Heisenberg himself drew the graph showing the rise in the number of neutrons from which it was possible to work out the point at which the pile's state would become critical – that is, when a self-sustaining reaction would begin which would need no injection of neutrons from any external source. The excitement was so intense that no one troubled about safety precautions, no one even paused to wonder whether their piece of cadmium would save them if a critical state was reached. But it was not reached. The quantity of uranium and heavy water proved to be insufficient.

Heisenberg calculated that they needed another 750 kg of deuterium and about the same amount of uranium. Meanwhile, 200 kilometres away at Stadthilm, Diebner had far more uranium and heavy water than Heisenberg lacked. In these closing days of the war

Heisenberg made a supreme effort to get hold of Diebner's uranium. But it was already too late. All that the German scientists managed to do during the death agony of the Reich was to hide the uranium and heavy water – and even that proved futile.

Almost all the leading German atom scientists were interned. They were accommodated in a luxurious country house near London. They were provided with tennis courts, newspapers and radio. Their demeanour was calm and arrogant. They were proud of the fact that the secrets of uranium were known only to them, that they could do something which no one else on Earth could even come near to doing. And then on 6 August 1945 they heard the news about Hiroshima. Their first reaction was bewilderment, then they decided that this was just propaganda, that this was simply a new kind of bomb which had nothing to do with uranium.

However, when the full details were broadcast that evening they realized that this was indeed an atomic bomb and became hysterical. They started to accuse each other of not working hard enough, of organizing things badly. The question that devastated them was: "How were they to go on living after such a blow to the prestige of German science?" [41, p. 111] It was only afterwards, 24 hours later, that they shifted to a different tack.

The following day Heisenberg had everything clear in his own mind and introduced a new theme to his colleagues: they had known and understood everything but they could not permit such an awesome weapon to fall into the hands of Hitler. They had built their reactor for peaceful purposes, they had intended to generate electricity.

Rumer (left) and Lev Landau (under cosmic showers)

Chapter 12

Cascades and Showers

At the end of September 1932 Yuri Borisovich Rumer returned to
Moscow. He had been away for five years. How Moscow had changed
in that time! How everything in Russia had changed! In the air there
was a smell of freshly-sawn pine and birch planks mixed with the smell
of whitewash. The city was one huge building site; there was wooden
scaffolding everywhere. Alleyways and sometimes whole streets had
been swept away. Even a native Muscovite could easily get lost in the
very centre of the Moscow of the 1930s.

The new buildings that were going up every few metres looked
from a distance like gigantic matchstick houses made of gigantic
matchstick logs. Closer to, however, it was possible to make out young
girls in red head-scarves and lads with long forelocks running – without ever stopping, it seemed – up and down hastily but sturdily built
wooden ladders. This unceasing activity made the new buildings look
like antheaps.

Despite the difficulties of day-to-day living (the rationing system
was still in force) the city hummed with enthusiasm and good humour.
The slogans of the times put into words the various aspects of a programme which all were striving to complete in the shortest possible time.
If the slogan was "We shall complete the five-year plan in four years!"
that is just what happened – unless, of course, it was completed in two
and a half. Another slogan was "We shall remove the need for imports"
and the need was indeed removed. Not only were imports of consumer
goods brought to a halt, the USSR also began to produce its own

machine tools, precision instruments, turbines and heavy self-propelled machines. During the first five-year plan just the "Leningrad" district of Moscow alone reported that twenty new factories had been commissioned. [43, p. 299] That is how fast things were moving in those days.

Moreover, each citizen felt a sense of ownership along with the conviction that the fate of his country depended on him and on what he did. "All 420 minutes of working time for production!" – working time was counted, not in hours, but in minutes.

The University of Moscow welcomed Rumer in a warm and business-like manner. He was immediately offered a teaching post in the department of theoretical physics and dragged off to a lecture. There was a notice in the corridor: "Who wants to be a professor?" and an arrow "pointing to the box where applications were to be deposited." [23]

Still groggy after his long journey, Rumer listened in wonder to a lecture about the latest advances in the study of atomic nuclei given by Professor Fowler of Cambridge. This lecture was delivered on 24 September 1932.

"This present year 1932," began Fowler "has turned out to be an annus mirabilis (year of miracles) for physics. A short chronological account of the harvest of new discoveries gathered in during the year may be of interest."

One great step forward made in 1932 was the discovery of the neutron and a study of some of its properties. Fowler spoke about Chadwick's experiments, about the mass of the neutron according to Chadwick's calculations and about certain predictions based on this discovery.

Chadwick had continued the extremely useful observations made by Joliot and his wife Irène Curie – while they were studying the penetrating radiation emitted by beryllium when bombarded with alpha-rays. In addition, Chadwick was able to prove beyond reasonable doubt that at least part of this penetrating radiation consisted of particles which had mass I, charge O and kinetic energy equal to 4×10^6 eV. He called these particles "neutrons". [44, p. 37]

The second great discovery of 1932, according to Fowler, was that made by Cockcroft and Walton, working in Rutherford's laboratory. Intensive study of the atomic nucleus had begun in 1919, when Rutherford had achieved the first artificial nuclear reaction by bombarding nitrogen with alpha-particles and so turning it into oxygen. In Rutherford's experiments, as in all others carried out with the aim of transforming nuclei, natural radioactive elements had provided the "shells" for the bombardment. For more advanced research, however, the energy of these natural "shells" was insufficient. The search therefore began for artificially accelerated charged particles. The 30s saw the birth of the accelerator. Almost simultaneously in three different places three accelerators of totally dissimilar design were assembled. The fact that the creators of these accelerators belonged to the quantum generation comes as no surprise.

The first accelerator – the electrostatic generator – was built in Princeton in 1931 by the American physicist Van de Graaff (born 1901). In the same year, at the University of California in Berkeley, Ernest Orlando Lawrence (born 1901) invented an accelerator based on the resonance acceleration of particles. In this device the charged particles pass repeatedly through an accelerator gap in a magnetic field, gaining great energy even under a modest voltage. Lawrence gave the name "cyclotron" to his accelerator. The first cyclotron was small enough to fit on the palm of his hand. The development of accelerators right up to the most modern types has followed the principles of Lawrence's cyclotron. Also in 1931 Cockcroft (born 1897) and Walton (born 1903), while working in Cambridge under Rutherford, managed to build a successful accelerator of the cascade generator type, which works by multiplying the voltage obtained in the separate cascades of the system. This accelerator was the first to produce really striking results. Cockcroft and Walton obtained their first atomic reaction with it in 1932 when they split an atom of lithium using accelerated protons. This was what Fowler called the second great achievement of the year.

"After these astounding advances all the rest of the events which I must mention may seem humdrum and boring, but nevertheless they are all part of the very real success that has been achieved in our effort to reveal the nature of the atom", Fowler went on. [44, p. 39]

Rumer was not astounded, however, by the advances of which Fowler was speaking. In fact, he had heard about them while still in Göttingen. On the eve of his departure he had taken an active part in a discussion of the latest news, which had not yet appeared in print but had arrived by way of what is now called "private communication". He had been most impressed by two things: one was Heisenberg's new hypothesis of the structure of the atomic nucleus, which he had formulated straight after learning of Chadwick's suggestion that the nucleus is made up of protons and neutrons; secondly, Anderson's discovery of the positron – i.e. of that same non-identical twin of the electron, the particle with the same mass and spin as the electron but with a positive charge, that Dirac had predicted.

Rumer was also impressed by something else – the very fact that this Cambridge professor went straight on to talk about the latest scientific achievements of his own University of Moscow. Rumer could not help recalling that in this very same university ten years previously, when he had been taking his final examination in physics and he had written Maxwell's equation in vector form, the lecturer had looked at him quizzically and said: "Of course, I understand what you have written here – but how do you know about this expression? Just write it out as it is, if that's all right with you – all nine of Maxwell's equations in component form."

The teaching of physics at Moscow University, and even more the scientific research done there, were in a lamentable state for a long time after the revolution. In this and other fields (the exception being mathematics) Moscow University found it difficult to recover from the assault made on it in 1910 by the Tsar's minister Kasso. 138 of the best professors and lecturers were then forced to leave their posts. Among them went the founder of the first Russian school of physics – Pyotr

Nikolayevich Lebedev. More than a thousand students were expelled at the same time. The school of physics, which had only just been born, was destroyed. It was really only at the end of the 1920s, with the arrival of Leonid Isaakovich Mandelstam and Sergei Ivanovich Vavilov, that Moscow physics began to revive. Its resurgence was not a gradual one; it happened at a breath-taking pace. Literally within two or three years Moscow's physicists were working right at the cutting edge of world science.

It suffices to say, moving on from Fowler's list of discoveries, that a few months later Cockcroft's and Walton's result was also obtained in Kharkov by Walter, Sinelnikov, Leipunsky and Latyshev (then known as "the boys" – they were all younger than the century). In 1937 Walter and Sinelnikov were to build the biggest cyclotron in Europe. Moreover, Dmitrii Ivanenko had put forward his hypothesis of the proton-neutron structure of the atomic nucleus independently of and, in fact, slightly earlier than Heisenberg. Ivanenko's article was already in print (in "Nature") in October 1932. Heisenberg's article on the same topic was published later.

On the first of October 1932 Yuri Borisovich Rumer was appointed as a lecturer in the department of theoretical physics and on 15 January 1933 he became a professor of Moscow University.

Those first months seemed to be totally taken up with meeting old friends and telling them all about what had happened in those five years spent in Germany, about the astounding achievements in physics which he himself had witnessed, about the Nazi plague and the horrifying prospects arising from it. And, naturally, there were lectures to deliver, several courses of them all at once. Rumer's lectures soon became popular. Attracted by the lecturer's brilliance, by his mastery of the mathematical framework and by his knowledge of all the very latest developments in physics, even non-physicists – mathematicians, chemists and linguists – came to listen to him. It was not only the latest discoveries themselves that he put over to his listeners in the clearest and most accessible form. He also managed to re-create the very atmosphere in which the discoveries were made. Rumer was in demand everywhere. Apart from his normal university courses he had to give popular lectures in the oddest places. He always enjoyed such events. Life was in fall spate – and he was home again.

Meanwhile in Germany his papers were being published one by one: "Towards a Theory of Spin Valency" (Nachr. Ges. Wiss. Göttingen, 1932, No. 4); "A Basis for Independent Invariants in Vector Space" (ibid. No. 5) written in collaboration with Hermann Weyl and Edward Teller; "A General Theory of Transformations in Hilbert Space" ("Nature" 1932); finally, "Concerning Invariants of Hilbert Space" (Phys. Zeitschr. Sov. Un.) published in German in a Soviet journal.

In the Soviet Union, as in other countries, scientific articles were still published in German – it was then unthinkable that anyone should work in science without a knowledge of the language of Planck and Einstein. However, these were the closing moments of that era in which science spoke German. Fleeing from Hitler's regime, scientists were leaving Germany and seeking refuge in Britain, Ireland and France. Many of them came to the Soviet Union. In 1933 a large number of first-class scientists found their way to Leningrad, Kharkov and Moscow. They left Germany and threw in their lot with Russia – some only for a time, others for the rest of their lives.

Among the physicists were some of Yuri Borisovich's Göttingen friends. Hans Helman, with whom Yuri Borisovich had tried to get into the Nazi rally, turned up in Leningrad. Viktor Weisskopf went to Kharkov.

Kharkov exerted the strongest pull at that time – especially on the theoretical physicists. "Kharkov in the 30s meant Landau", recalled Yuri Borisovich. And the fact that the students were, in many cases, older than their teacher did not worry anyone. Landau created a brilliant school of theoretical physics. Many people went there hoping to stay at least for a little while and perhaps, with luck, for good. Viktor Weisskopf, the future founder of CERN, wrote: "I could not find work in Germany, England or France. In 1933 I went to Russia and stayed almost a year in

Kharkov. Landau, Lifshitz, Akhiezer and many other young Russian physicists were working in Kharkov at that time. Life in Russia was far from easy but it was interesting and instructive." [35, p. 22]

In 1935 Fritz Houtermans appeared in Kharkov. Hunted by the Nazis in his own country, Houtermans had managed to escape to Russia. He worked in Kharkov from 1935 to 1937. He did some very interesting research which led to some important findings. He worked with Kurchatov on those occasions when Kurchatov came from Leningrad on visits to the physics institute in Kharkov. Houtermans' unbounded imagination gave birth to some bold ideas. It had been he, after all, who, while he was still in Germany, had been the first to point out that thermo-nuclear reactions are the source of the energy which lights the stars. In Kharkov he carried on with his investigations into the possibility of creating thermo-nuclear energy sources on Earth. However, as history was to show, the world was not ready for Houtermans' ideas about thermo-nuclear reactors or for his tentative calculations. Their time had not yet come. As often happens, the questions he raised eventually received more serious attention in a different place and time.

Everyone in Kharkov liked Houtermans – he was the life and soul of any gathering. People watched over him and looked after him, sympathising with him as an exile and a victim. When he was suddenly deported to Germany in 1937 his Russian friends were horrified and baffled. They tried not even to talk about the fate that awaited him there – it was all too obvious. But, as we already know, fate was to bring Fritz Houtermans back to them during the war, when the Germans were on the rampage, robbing all the institutes of Europe and carting off scientific equipment to Germany.

When East-West relations warmed a little in the 1960s and Soviet scientists began to make trips abroad Houtermans tried to meet them, especially those whom he knew personally. At these meetings he invariably complained about the way things had worked out for him. He had lived a long time, he could have done a lot, he had loved his friends – and yet he was a stranger wherever he went. He always maintained that his heart belonged to Russia.

Kharkov in the 30s opened its doors to the foreign fugitives while also nurturing the young people who flocked there from every corner of the Soviet Union. Conferences and symposia were constantly being arranged in Kharkov. Niels Bohr, Pauli, Dirac, Fowler and many other famous physicists attended them. Kurchatov regularly came to Kharkov on long visits which sometimes lasted a month or more. His young assistants used to come with him. From 1933 onwards research in nuclear physics was an important part of Kurchatov's programme.

Yuri Borisovich Rumer, too, was a frequent visitor to Kharkov. It was there that his friendship with Landau was finally cemented and where their scientific collaboration began. This was a wonderful, almost carefree time when questions of politics, whether domestic or foreign, hardly troubled them.

Nor did they allow the practical difficulties of everyday life to depress them. Their days and nights were filled with art, poetry, endless scientific debate and hard work. Naturally not all the debates were about scientific matters. Desperate arguments could flare up over anything under the sun.

With Gamov's arrival from Leningrad came a subtle change of key in Kharkov. Gamov was a jovial character who enjoyed mocking others but was both kind and intelligent. However, since his return from abroad he seemed somehow to have changed. Sometimes it was difficult to tell whether he was joking or serious. He would, for example, say very seriously and with apparently genuine melancholy that he was "fed up" and that he could live like a human being only "there". He said that, although all his cunning attempts to get out of the country had met with failure so far, he would nevertheless one day think up something really clever. Later he did indeed think something up but there was nothing very clever about it when it finally happened – he simply never returned from a trip to Britain. At that time, however, nobody believed him and people did not encourage him to enlarge too much on his misgivings about the alarming events happening in the Soviet Union – although some of the things that they heard were indeed alarming.

When Kapitza came to Moscow from England in 1934 after working with Rutherford for 13 years it was suggested to him that he stay in the USSR and found an institute which he could shape to suit himself.

Kapitza agreed. The news of this great event quickly reached all the various communities of physicists. Rumer did not know Kapitza personally at that time and had no way of knowing the full details of the case. As for the rumour that the authorities had simply refused to let Kapitza go back and that he was being compelled to stay – he dismissed it as unfounded. After all, Rumer himself had just returned and was happy to be home.

However, Pyotr Leonidovich Kapitza wrote at this time to V.I. Mezhlauk, who was the deputy chairman of Sovnarkom. Valerii Ivanovich Mezhlauk was one of four Mezhlauk brothers, all of whom served the revolution. Martin, the youngest, was a 23-year old provincial commissar for justice when he was shot by White Guards in 1918. The eldest, Ivan Ivanovich Mezhlauk, became a member of the Central Executive Committee but was arrested and shot in 1938. V.I. Mezhlauk suffered a similar fate, also in 1938. What happened to the fourth brother is not known. Here is an extract from one of P.L. Kapitza's letters which were presented to the author by P.E. Rubinin:

"Leningrad 2 November 1934, Comrade V.I. Mezhlauk!

In answer to your communication of 26 October (ref. no. 29/C.M.), which reached me only in the evening of 31 October and in which you suggest I inform you of the scientific work which I propose to do in the USSR, I can give the following details ..."

Pyotr Leonidovich went on to describe how the work he had done in Cambridge had developed over 13 years and how the co-workers whom he had so carefully selected had also developed as a team. He wrote about his repeated suggestions that young Soviet physicists should be sent over to him for training for, as he had pointed out to those in authority, this was the only way of transferring his work to the Soviet

Union. "However", he wrote, "To my deep regret, this was not done."
He now saw no possibility of continuing this work in the Soviet Union
and had therefore decided to turn his attention to research in biophysics.
He had already come to an agreement with Ivan Petrovich Pavlov, who
had offered him a post in his own laboratory.

In P. Kapitza's archive there is a "memorandum" written by Paul
Dirac in which there is specific reference to the presumed grounds for
Kapitza's detention: "Three reasons for detention.

(a)   an unfounded allegation sent from England about military
       research;
(b)   Gamov wrote to Molotov asking to be granted the same status that
       "K" (Kapitza) had up to October 1934;
(c)   knowledge that could be valuable in time of war.

He is most indignant that these three things should count against
him and has offered his resignation..."

Next come details of a "Plan for an amicable solution" – "course
of action that he considers correct", and the conditions which Kapitza
was presenting to the Soviet government.

In that first cool, restrained letter to Mezhlauk, Pyotr Leonidovich
does not make any request for cooperation in reviving his Cambridge
project. He already regarded that as a lost cause. However, as early as
December 1934 he wrote to his wife:

"... A lot of interesting things have happened in the last few days.
Yesterday evening I went to see V.I. (Mezhlauk) and we talked for
almost three hours. These were some of the most delightful hours I
have known in recent weeks. V.I. is an intelligent man and understands
what you mean before you've finished speaking, although sometimes
what he says and what he thinks are two different things – but this
must become an obligation for a man in high office ... Anyway, I have
no doubt that V.I. and myself can eventually see eye to eye and since he
and I share the same desire – to strengthen Soviet science – I would be
glad of the chance to work with him in the future ..."

They would indeed work together later. But not for long. The optimistic tone of Pyotr Leonidovich's letter to Anna Alekseyevna about his meeting with Mezhlauk (written when Kapitza had been held in the Soviet Union against his will for about two months) should not be taken to mean that his problems were over, that everything had been sorted out. Far from it. He wrote to Molotov about this period of his life: "For the first four months nobody paid any attention to me and I could not even get bread coupons. Then for three months, obviously just to scare me, two NKVD agents walked alongside me everywhere I went, now and then amusing themselves by coming up close and tugging at my coat ..."

To Stalin he wrote: "... When I was suddenly detained more than a year ago my research was interrupted at a most interesting point and I suffered a lot. Then I was subjected to some really disgusting abuse. These months back in the Soviet Union have been the most dreadful I have ever known. Even though I can see the sense of transferring my work here, I certainly cannot understand why it was necessary to treat me so cruelly.

Apparently I was suspected of having done something wrong – nobody ever told me what. The attitude towards me was malevolent and suspicious in the extreme. Whatever I said was twisted around to mean something else ... I have to say bluntly that the attitude shown towards me is not such as will enable scientists to value themselves or their work – in fact, it will destroy their enthusiasm ..." This assertion of the importance of enthusiasm is a theme secondary to the main business of the letter.

The letter was written for two main reasons. Firstly, it was a response to the news that the senate of Cambridge University had finally approved the transfer of Kapitza's laboratory to the Soviet Union (the letter is dated 1 December 1935). Secondly, it is concerned with the difficulties involved in the building of the new institute. Pyotr Leonidovich writes of the institute's material infrastructure, about the need for "scientific housekeeping".

As has been noted, the secondary theme of the letter is the importance of enthusiasm in science and how essential it is to infect the rising generation with this enthusiasm. The end of the letter is simple, without any bowing or scraping:

> "... In conclusion I wish to say that however bad things get, however I am treated, I shall work flat out. I shall do my best to ensure that my work is successful, I shall fight for it to the end. The outlook is not bright but the only thing I fear is that my strength may not be sufficient – it may be that my energy will be wasted on troubles and trifles and I shall have none left for my work.
>
> P. Kapitza."

As we can see, Pyotr Leonidovich's problems were still with him a year later. They were to last even longer than that. However, that first meeting and first acquaintance with Mezhlauk gave him hope. Being a man of unusual perspicacity, able to make an absolutely sober assessment of the situation, Pyotr Leonidovich realised that it would be Valerii Ivanovich Mezhlauk, if anyone, who might support him in the difficulties which he now had to handle. Within two weeks of their first meeting (3 January 1935, Moscow, Hotel "Metropol" Room No. 485) Pyotr Leonidovich wrote to Anna Alekseyevna:

> "Dear Anya, today I was very glad to see the announcement in "Izvestiya" about my appointment as Director of the Institute for Physical Problems ... I am pleased that I can be of use in that post at least."

Building work started on the site of the institute – Lenin Hills – in January 1935. Pyotr Leonidovich had chosen the spot himself. The design of the institute along with every other aspect of it including its financial system, was his own creation. Difficulties beset him on all sides but for Pyotr Leonidovich everything connected with the job in hand, from the unusual financial arrangements already mentioned to the quality of the plastering, was important and serious. For him, nothing connected with his work could be trivial. This is what motivated his fanatically bold struggle to get things done.

He was quite capable of writing a mercilessly frank letter about precisely such matters as the plastering to Molotov or Stalin. He wrote, for example, to Molotov about problems with the building and the way the whole job was organised:

"So, you see, Vyacheslav Mikhailovich, just how many things have gone wrong. What conclusions do you draw from this? Personally, I think that the root of the problem is the awkward relationship that has always existed between you and me. In any case, one way or another the job has to be finished... If you had made an effort to involve me in the life of the country, which is more wonderful than you yourself think, then we would probably have been friends long ago..."

More often than not, however, appeals to Molotov fell on deaf ears. Letters to Stalin sometimes produced results; letters to Mezhlauk always did. It was Valerii Ivanovich Mezhlauk who championed Pyotr Leonidovich's cause. To him Kapitza wrote not only to get things done but also to express his anxieties and to send seasonal greetings. Especially noteworthy is Kapitza's New Year letter, sent to mark the beginning of 1937. It takes the form of a report on the work done in the year then ending and runs to many pages:

"Much-respected Valerii Ivanovich!

I want to write you some sort of report about this last year. When you look back and see where you have come from you know better how to go forward and after all (These two words are in English in the original – Translator) we have to some extent been working together, striving towards the same ends. Apart from you, no one in the Soviet Union has watched the progress of my work or knows about it in any detail. The main thing is that the institute and the living quarters are now finished ..."

What follows is a long and detailed account of all the work done in the institute, about past alarms and future plans, about finances and accounting procedures, about the five electricians, the eight mechanics, the two joiners and the one glass blower with whom Pyotr Leonidovich was particularly pleased and who, in his opinion, should form the permanent team around which the laboratory should function – while research workers could and should pass through the institute, first training there and then taking their acquired expertise to other establishments. At the end of the letter Pyotr Leonidovich writes:

> "And so, 1937 is beginning. Will you carry on helping me as before? Without your help we shall get nowhere and yet I realize that it is wrong to make such demands; your main responsibilities are extremely important and concern the whole country. I know I frequently have to bother you with trivial problems caused by the muddle which affects our society – but I do this only because I am sure that through you the experience gained in our institute will be passed on to others and will speed and strengthen the development of Soviet science...
>
> Happy New Year! P. Kapitza"

But Mezhlauk's fate was already sealed. Kapitza's last letter to Valerii Ivanovich (dated 19 November 1937) is quite short and concerns the dispute between Lysenko and Vavilov. In characteristic manner, combining simplicity of exposition with wisdom and complete objectivity, Pyotr Leonidovich writes about the laws governing scientific disputes and about how they should not be conducted:

> "Not only absurd but also harmful methods are beginning to be used in our debates. This tendency is not only to be observed in the quarrels of the geneticists .... Reduced to its simplest terms the argument goes like this: if in the field of biology you are not a Darwinist, if in Physics you are not a Materialist, if in history you are not a Marxist – then you are an enemy of the people.
>
> That kind of logic serves only to terminate fruitful discussion with 99% of all scientists. The use of such methods is not only harmful to science but also discredits the immense theoretical achievements of Darwinism, Materialism and Marxism. The combatants must be told in no uncertain terms: in your debates rely on your own intellectual powers, not on the power of Comrade Yezhov...
>
> Greetings, P.Kapitza."

However, in 1934–35 the stories about Kapitza sounded like mere gossip. Yuri Borisovich Rumer, however, who by then was constantly hearing alarming tales – and not only about Kapitza – was still a devout believer in the justice and rightness of all that was done in his country and could not believe that anything evil was happening.

"In 1934 or 1935", recalled Yuri Borisovich, "Gorky organised a publishing house which was to specialize in the memoirs of ordinary Soviet people. The staff consisted of well-known writers, artists and also "interviewers". The latter were women chosen for their pleasant personalities who were sent to talk to the people whose reminiscences the publishers thought it necessary to record. For some reason I, too, became mixed up in this business. The writer Fedin was appointed chief editor of my memoirs and a most attractive "interviewer" was allotted to me. I remember our first chat very well.

"Yuri Borisovich, let's start with the main thing – how did you become a professor?"

"Well, let me think. I was selected and given the post. I was recommended."

"And who recommended you?"

"My foreign referee was Schrödinger and here at home it was Mandelstam", I remember that this made no impression on her at all. However, she became interested when I started to tell her about Germany. Even I was happy with the result when Fedin reworked all this material and sent me a copy of the final version. Fedin knew how to write and he was pleased with me. Then, abruptly, this project was dropped. It was as if it had been blown away by a sudden gust of wind.

A short time later I saw a caricature of myself in the university newspaper. In the picture I was wearing knee-high socks and a sheepskin coat and carrying a small suitcase covered with foreign stickers. I was somewhat baffled. I thought that it must have been intended to be funny – and yet it was not funny. Mila said to me: "They'll put you in prison." But she always had a vivid imagination. It was ridiculous to believe that – and Yuri Borisovich still did not believe it, just as he had not believed it when a storm had broken in his vicinity long before the caricature in the wall newspaper.

The person at the centre of that earlier storm was Nikolai Nikolayevich Luzin, who was declared a saboteur and an enemy of Soviet science. This denunciation threw Rumer's mathematician friends into a state of bewildered helplessness which also affected Yuri Borisovich himself. In his heart of hearts he was convinced that this was all due to some kind of misunderstanding and that it would be cleared up. However, the "Luzin Affair" continued to gather momentum.

On 2 July 1936 Pravda printed a devastating article against Luzin – "A Reply to Academician Luzin". The attack was renewed the following day. The article of 3 July was entitled "Enemies in Soviet Disguise". That same day, as if they had been waiting for this article to appear, the scientific staff of the Mathematics Institute of the USSR Academy of Sciences held a meeting. The participants discussed the two "Pravda" articles and adopted a resolution. The resolution consisted of eight points, each of which was a sign of the times, a typical mark of the epoch then about to get into its stride.

How did it happen that phrases which defied ordinary common sense, which would not bear any kind of logical scrutiny, were not merely accepted passively by our society but were legitimized and became the norm? An epidemic of accusations broke out and all levels of society were infected – neighbour destroyed neighbour, colleague destroyed colleague, stable-boy destroyed veterinary surgeon.

Sadly, we can be sure that the leaders of the country who were responsible for those dreadful days hardly regarded Luzin as a personal enemy – he was as much an enemy, for example, as that 22-year old vet from a remote Siberian settlement or all the other tens and hundreds of thousands. People were invited to play a terrible game – and they played it. Many were the victims in this game but the culprits were probably no fewer in number. It is possible to imagine something really monstrous – that there was a target set for the number of enemies of the people handed in. If there was, then the country easily overfulfilled its quota.

In the end Luzin was defended successfully, although the persecution was to go on for many years. But no one cared about those other hundreds of thousands, except their own close friends and families.

The eight points of the resolution adopted at the general meeting at the Mathematics Institute were as follows:

(1) The mathematical community has known for a number of years the facts about the "activities" of N. Luzin which have now been brought to light in the pages of "Pravda". Moreover, the facts as published are an understatement.

(2) However, the scientific community failed to detect in these facts the face of an enemy disguised as a Soviet academician, explaining them as "eccentricities" in N. Luzin's character.

(3) We must, therefore, admit openly that to adopt such an attitude towards N. Luzin was to adopt a position of putrid liberalism, which provided favourable conditions for Luzin's foul anti-Soviet activity and made it easier.

(4) The magnificent Bolshevik vigilance which helped "Pravda" to unmask this enemy who managed to find his way into the ranks of Soviet scientists will serve in our future work as an object-esson in the struggle for Soviet Socialist science.

(5) We call the whole scientific community of our country to implacable struggle against the enemies of the people under whatever disguise they seek to hide. We call on our colleagues to adopt Bolshevik self-criticism as regards their work for this is the indispensable condition for the realization both of the highest potential of our scientific development and of the fullest possible unity between science and the practical building of socialism.

(6) The meeting puts to the presidium of the Academy of Sciences a proposal that N.Luzin be removed forthwith from the posts of Chairman of the Mathematical group of the Academy of Sciences and Chairman of the Mathematical Qualifications Commission.

(7) The meeting also asks the presidium of the Academy of Sciences to consider, in accordance with Article 24 of the Academy Regulations, the question as to whether N. Luzin can continue to be accepted as a full member of the Academy.

(8) The meeting considers that, in order to ensure the proper direction of Soviet mathematical life, it is essential to reinforce the mathematics group by bringing in new full and corresponding members" [45, p. 275]

This last point strikes, as they say, closer to the real target.

The wave of meetings and resolutions concerning Luzin rolled on to many other institutions. On 5 August 1938 the presidium of the Academy of Sciences brought out an announcement which began thus: "The victory of Soviet power and the immense success of Socialist construction have raised our country in an incredibly short space of time from unparalleled disorder and age-old inertia to the level of a first-rank world power. The working class of the whole world looks to the Land of the Soviets with hope and expectation, for it rightly sees in her its main citadel in the struggle for higher forms of human culture" ... [45, p. 277] and so on about our difficult path, about the real achievements of our science and, of course, about the enemies of the people, who stand in the way of progress.

"... The Presidium of the Academy of Sciences confirms that the widest circles of Soviet society have participated in the discussion of the Luzin affair and that there has been unanimous condemnation of Luzin's anti-Soviet activity, his hypocrisy and his deceitful behaviour... the Presidium of the Academy of Sciences is of the opinion that the behaviour of Academician N.N. Luzin is incompatible with the honour of full membership of the Academy of Sciences and that our scientific community has every justification for considering his exclusion from the roll of academicians.

However, taking into account Luzin's importance as a leading mathematician, as well as the full force of public opinion which manifested itself in such widespread, unanimous and just condemnation of

N.N. Luzin, and acting upon a desire to offer Luzin a chance to remodel both his conduct and his work in the future, the Presidium considers that it is sufficient to warn N.N. Luzin that if there is no decisive change in his conduct, the Presidium will without delay raise the question of his removal from the list of academicians" [45, pp. 277–278].

This material was published at the end of one of the issues of the journal "Achievements of the Mathematical Sciences". The first page of this journal (and the second, since there was insufficient space on the first) was devoted to a leader article entitled "Eliminating Luzinism from Soviet Science", which mercilessly "exposed" Luzin and proposed eliminating not only Luzinism but also those attitudes, extraordinarily harmful in present conditions, which can be lumped together under the heading of "academic traditions", which rule out a direct and hard-headed assessment of the concrete deficiencies in the work of particular academics ..." [45, p. 4]

It is hard to imagine the scale of the damage done by the full text of this announcement made by the Presidium of the Academy of Sciences. The main point, however, lies in its last paragraph – that is, the decision to leave Luzin alone. Those who were able to pride themselves on their "magnificent Bolshevik watchfulness" were not satisfied. Now they started to terrorise not only Luzin but his friends and pupils, too. This went on for a long time. None of Luzin's pupils could bear to recall that period. Only once during my enquiries about Schnirelman did Lusternik happen to mention this subject. "Nobody could believe that Lev was capable of it. He was such a carefree, fun-loving sort and suddenly he committed suicide. He had been called in for questioning. He came home in the evening, shut himself in the kitchen and turned on the gas."

Luzin was saved. At least, he was saved from prison – he was not saved from humiliation and persecution. Despite all this, Nikolai Nikolayevich Luzin made no attempt to "remodel his conduct and his work". He continued to be himself and to work as he had always worked. The only difference now was that his young followers met like conspirators at his home or at his dacha to pursue their late-night – sometimes all-night – mathematical studies.

Luzin died with extreme suddenness in 1950 from a massive heart attack. There was no obituary but an old article of Luzin's was published in one of the 1951 issues of the journal "Achievement of the Mathematical Sciences". There followed a long article written by Nina Bari (that same Nina Bari who had divided her grant into tenths back in 1920) and Lazar Lusternik. The title of this article was "N.N. Luzin's Work on the Matrix Theory of Functions". A short footnote read: "... the editors plan to publish further material relating to N.N. Luzin's work in a forthcoming issue."

The next issue did not appear until 1952. In it there was once again one of Luzin's old articles followed by two long articles which examined Luzin's contributions to two distinct areas of mathematics. The same pattern was repeated the following year but, ironically, the opening pages of the issue in which the last instalment of the Luzin series appeared were taken up by a black-bordered portrait of the recently deceased Stalin and the text of the Soviet Government's message to all the workers of the Soviet Union. To the first of the articles dedicated to Luzin was added a footnote which, for the first time, mentioned his death: "To mark the third anniversary of the death of N.N. Luzin the editors here publish three articles which, together with those published in the previous two issues, form a survey of Luzin's mathematical thought".

By the time this series of articles was published Yuri Borisovich Rumer had served his sentence and the most difficult period of his life was beginning. However, back in 1936 the treatment of Luzin seemed to him to have arisen from some monstrous misunderstanding.

In 1937 Landau moved from Kharkov to Moscow to work with Pyotr Leonidovich Kapitza. Kapitza had invited him to head the theoretical section of his institute, the construction of which was then nearing completion. Events almost took a rather different turn. A year previously a different candidate had been considered for the post. This candidate was none other than Max Born.

Max Born's friendship with the Kapitza family had begun in 1933 when the Borns had been forced to leave Nazi Germany and seek temporary refuge in England. From 1935 to 1936 Born worked at the Bangalore Institute in India. From there he wrote to Anna Alekseyevna Kapitza: "When I read the British and German newspapers I sense that Europe is moving towards a new catastrophe. I have already lived through quite enough "historic events" for one lifetime. It seems to me that Russia is the only country that is ruled to any extent by reason. If I were not so old and so hopelessly spoiled by bourgeois habits I would ask Kapitza to find me a job in Russia. We can discuss this matter when I get back to Cambridge in April ..."

This was enough to prompt Kapitza to write to Mezhlauk about Max Born, saying that there was "a chance to win this scientist for the Soviet Union". In his letter of 19 February 1936 he argued:

"... If we secure him

(1) The Soviet Union will have gained a first-class scientist.
(2) We shall have obtained for our young people the leading teacher of theoretical physics.
(3) I would like to give him a post in our institute; his presence would give our work a powerful boost."

On the 26 February Kapitza wrote to Born:

" Dear Born,

Your letter to Anna was delivered here and I took the liberty of reading it. I found it most interesting. I was sorry to hear that you have not yet decided which hemisphere you are going to settle in. You have my sympathy – everyone wants you and you are faced with such a wide choice that you cannot make up your mind. Perhaps I am more fortunate than you as I have no choice to make. I felt much happier when I heard about the decision to ship my apparatus. I now live in hope that in a few months from now I shall be able to take up my work again. It has been suspended now for almost two years. The other good thing is that Anna and the children are now with me.

Working conditions are far from ideal but will soon improve and everything is being done here to enable me to get on with my work. It was in Cambridge, a town with an ancient cultural tradition, that I reached maturity. I spent 13 years there! You, on the other hand, have spent the greater part of your life in Göttingen, which enjoys the same sort of intellectual life as Cambridge. It is, of course, pointless to judge the current state of affairs here in the Soviet Union by standards such as these.

However, it cannot be denied that it is extraordinarily fascinating to watch the growth of a new culture and of new principles and I do not regret the fact that I have appeared on the scene here, able to take an active part in this game. Of course, I do feel and will always feel resentment about the swinish way in which I have been treated but, my dear Born, no government of any age or of any part of the globe has ever been famous for its sensitivity and the individual who falls into the "maelstrom of history" is likely to be smashed. The only thing to do is to cultivate spiritual resilience ...

Your letter gave me the idea of playing a malevolent trick on you, the idea of adding to your anguish by suggesting you add our one-sixth of the Earth's surface to your list of possible destinations. I think you may well give this suggestion serious thought. Apart from arguments based on wider historical considerations there are other arguments that may win you over:

(1) Here you will be able to found your own school of theoretical physics;
(2) You will receive a rapturous greeting here and will never be made to feel that you are a foreigner* or that you are standing in anyone's way;
(3) our theoretical physics is not strongly developed and your leadership will be accepted most willingly.

Your bourgeois habits will, of course, suffer a little but our standard of living is rising very fast and I think that even now you would be quite satisfied with the comforts that are available. You might well feel that shopping is a burdensome business and that the road surfaces are poor but you would find some compensation in the theatre and the concert hall; our standards in these areas are higher than those of many other countries. You would be well provided with all kinds of literature. Altogether, your spiritual well-being would be greater than your physical comfort!

Give this idea some thought and write to me. Personally, I would be happy if you moved here and I would be pleased to have you on the staff of our institute as supervisor of our theoretical section. We could offer you a small five-room cottage in the grounds of the institute. It has gas, electricity, hot water and a heated garage. It stands in the grounds at the top of a hill

---

* Even Kapitsa, himself a victim of state harassment, could not foresee the day when nearly all the foreign experts who came to the assistance of the young Soviet state would end up behind bars. – *Author*

overlooking Moscow. You could choose your own students, give lectures in the university etc. We could hold joint seminars, as we did at Cambridge. You could take part in our laboratory's research programme. I have not yet appointed a theoretical physicist and would leave the choice to you if I knew you were intending to come.

Think about all this and let me know how you feel about it. If my offer makes sense to you I will think about how to arrange everything with the authorities. I have reason to believe that my idea will meet with approval and support ...

I await your reply in the hope that my suggestions may appeal to you. Meanwhile please accept cordial greetings and best wishes

Sincerely yours, P. Kapitza"

In September 1936 Born should have travelled to Moscow for the final clarification of this proposal. However, by this time a professorial chair had become vacant at Edinburgh University and the Borns moved to Scotland. Kapitza brought in Lev Landau as head of the institute's theoretical section. From then on and until he died Landau headed the theoretical physics section. This was where he created his famous school of theoretical physicists. In the 1960s a group of his pupils, headed by Isaak Markovich Khalatnikov, founded the Landau Institute of Theoretical Physics at Chernogolovka.

For Rumer, Landau's move to Moscow was a cause for celebration. Now they were together all the time. Between arriving from Kharkov and moving into the cottage in the grounds of the institute Landau stayed with Rumer. Rumer had a room in a crowded but happy communal flat at No.68 Gorky Street. The two friends found time for everything : visits to the theatre, parties, etc.

Some of the parties consisted of endless poetry readings. Landau loved Kipling's verse and regarded himself as a "Simonist". Rumer also loved Simonov and, indeed, poetry in general – even in old age he seemed to be able to recall every line of every good poem ever written. Sometimes the parties were enlivened by humorous sketches.

However, work was, to use a term from physics, their "ground state". Each of them had his own interests. This was the time when

Landau produced his phase transition theory and his statistical theory of the nucleus. He was working on the diffusion of X-rays by crystals. He also demonstrated that neon and carbon are stable with respect to the alphy decay.

Rumer was working on the theory of elementary particles, the theory of superconductivity. By this time Yuri Borisovich had published two monographs: "An Introduction to Wave Mechanics" and "Spin Analysis". The first of these two monographs was intended as an introductory text-book for students and serves its purpose admirably, being written in a clear and lucid style. The book "Spin Analysis", on the other hand, was the first Soviet exposition of the new mathematical method in physics and remained the only one for several decades.

There were also projects, of course, on which Landau and Rumer worked together. Two joint articles came out in 1937: "The Absorption of Sound in Solid Bodies" and "The Formation of Showers of Heavy Particles". In 1938 Landau and Rumer published "A Cascade Theory of Electron Showers". It was an enormously important work.

At that time cosmic rays were practically the only available source of elementary particle radiation. Cosmic rays were the natural particle accelerators in which the new elementary particles were first discovered. Moreover, man-made accelerators were then still many times less powerful than these natural accelerators.

However, the processes going on in the atmosphere, which is constantly bombarded by these energetic particles arriving from space, are extremely complex and there was an urgent need for the formulation of mathematical laws upon which could be based the proper interpretation of the data collected down on the Earth's surface. Landau and Rumer established the formulae governing the formation of cosmic showers. They determined the number of particles in a shower as a function of the depth of penetration into the atmosphere for any given initial energy of the original particle and the distribution of the energies of these particles at a given depth in the atmosphere. They also studied the behaviour of cosmic showers passing from one medium to another. Their conclusions were confirmed by a mass of experimental data and the paper served as a basic manual on cosmic showers.

This was also the time when they had the idea for a popular scientific book called "What is Relativity Theory?" Once conceived, the book was delivered quickly. Landau provided a jocular review: "Two rogues persuade a third that in exchange for a ten-kopek piece he can understand what relativity is all about." Landau also explained what he meant by this: "It is clear why Rumer and I are referred to as two rogues – but why is the reader the third rogue? He is a rogue just because he expects to understand relativity for as little as ten kopeks." The rest of the history of this remarkable book was not as light-hearted as its beginning. Although it was eventually translated into more than 20 languages it was not published until 1959 and then only thanks to Evgenii Mikhailovich Lifshitz, who preserved the only manuscript copy of this book written by two "enemies of the people".

Prison in Tomsk (sketch of Rumer)

Chapter 13

Pentoptics

In 1938 Landau and Rumer were arrested. They were arrested on the same day and put together in a cell. "We smiled at each other," said Yuri Borisovich, "and, of course, concluded that they wanted to listen in on our conversations and so had put us in the same "envelope" – a tiny cell in the Butyrki prison in which it was possible to stand or sit down but not to walk about. In fact, as we afterwards found out, our sharing of the cell was the result of mere routine incompetence. Actually, the order for my arrest had been issued on 26 April and I was not brought in until the 28th. They just had not had the time to come and get me – there were so many more like me to deal with.

The first thing that Dau said to me was typical of his sense of humour: 'A fine birthday present they've given you, Rumchik. Many happy returns!' "

28 April was Yuri Borisovich's birthday. Having expressed his birthday wishes – which Rumer received without much pleasure – Landau went on as if nothing untoward had happened to give an account of his latest scientific breakthrough: "Listen, Rum, you'll never guess what I have come up with! I have worked out the secret of liquid helium!"

In 1937 Pyotr Leonidovich Kapitza had discovered a paradoxical phenomenon: helium, cooled to super-low temperatures close to absolute zero, not only refuses to solidify, it actually loses viscosity and becomes super-fluid. Rumer was the first person to hear Landau's brilliant explanation of the superfluidity of helium. It is an amazing phenomenon, in which quantum effects on a massive scale are involved. Three more years were to pass before Landau's explanation became known to the world. 25 years were to pass before this explanation was marked by Nobel prize.

The pair spent only one night together before being escorted to different cells. Landau was to remain in prison for one year, Rumer for ten. This ten-year sentence was followed by deportation and exile. The two friends did not meet again for fifteen years. When they next met Landau was, as before, head of the Theoretical Physics Section; Rumer was teaching at the Yenisei Teacher Training Institute.

Rumer was not kept long at Butyrki. Yuri Borisovich recalled: "They made such wildly impossible accusations, "exposed" such ludicrous crimes, that I concluded that there was no point in objecting or defending myself. The main thing was to preserve my peace of mind and good humour. On the third day after my arrest they handed me a letter written by Landau to Yezhov and asked: "Do you know your friend's handwriting?" "Yes, I do," I answered. "Could you tell if someone tried to fake his handwriting?" I said that I could. "Just read what he has written, then." I read it: "To People's Commissar Yezhov. I, Professor Lev Davidovich Landau, hereby declare that I organised a group of physics professors, including Yu.B. Rumer, I.E. Tamm, M.A. Leontovich etc., the aim of which was to undermine theoretical physics in our country. For example, I myself, in my work on quantum electron gas, deliberately left out all information about practical applications and left in only bare formulae ... " etc.

"There you are," said my interrogator, "Are you convinced? And now let's agree that without any application of physical methods you will place your own signature on this very document." "What the hell!" I thought, "You only die once." and so, "without any application of physical methods", I signed the paper. That is probably why I was never once beaten. My wife still doesn't believe it. It is true that I was put in a "cupboard" where I could not sit down, that they shone bright lights in my eyes, insulted and taunted me – but I was not beaten.

I was sorry for those people. They really seemed to think they were up against a genuine spy and they did try to do their job properly. Many years later in Moscow University I ran into the man who interrogated me at Butyrki. I was once again a free citizen with all my rights restored. He was in civilian clothes. I would have walked past him but he stopped me: "Yuri Borisovich, I want you to know that a year after your arrest, when you were being sentenced, I submitted a protest to the military prosecutor saying that I had established that the only offence with which citizen Rumer who was arrested in April 1938, could be charged was that he had associated with Landau, an enemy of the people. Moreover, when Rumer came to be sentenced the said Landau had already been released by the organs of the Ministry of Internal Affairs. For this reason I, Prosecutor ... of the Soviet Army wish to protest against this sentence and ask for it to be rescinded."

By the time he sent his protest I was already working on aircraft wing flutter – partly, it would seem, thanks to this same interrogator. After I had signed Landau's "confession" the interrogation continued and on one occasion in the middle of all this nonsense I heard the question: "Are you prepared to work for the Soviet Union as an engineer?" "Of course!" was my delighted reply. That's how I ended up with the aeronautical engineers at Bolshevo."

Landau immediately got into much worse trouble, bringing down upon himself the fury of the security organs. Convinced that everything would soon be cleared up, he maintained from the start an independent attitude and let them see how disgusted he was by what was happening. It was only later on, in order to avoid being beaten, that he started signing interrogation records which contained absurd confessions. Landau was doomed – and would have perished but for Pyotr Leonidovich Kapitza. On the day Landau was arrested Kapitza wrote a letter to Stalin:

"28 April 1938, Moscow

Comrade Stalin!

This morning a member of the Institute's Scientific staff, L.D. Landau, was arrested. Although he is only twenty-nine years old, he and Fock are our greatest theoretical physicists ...

... There is no doubt that the loss of Landau and his work could not pass unnoticed and would be keenly felt in our institute, in Soviet science and indeed in the international scientific community. Of course, learning and talent, however considerable, do not give a man the right to break the laws of his own country and if Landau is guilty he must answer for his crimes. However, I beg you, in view of his exceptional ability, to order that his case should be handled with the greatest of care.

I think I should also give an assessment of Landau's character. Bluntly speaking, his character is dreadful. He is a quarrelsome bully who loves to seek out the mistakes of others – especially those of important senior colleagues such as our academicians. When he spots such errors he mocks those who make them, showing a disgraceful lack of respect. In this way he has made himself many enemies.

... but despite all his character faults I find it difficult to believe that Landau is capable of doing anything dishonest.

Landau is young, he still has much to do in science. No one can write about this with such certainty as can another scientist, which is why I am writing to you.

P. Kapitza"

Kapitza's letter requires no commentary – except perhaps a note about the mention of Fock in the first paragraph. It is, of course, not by accident that his name appears here – indeed, every word in the letter is carefully chosen.

Vladimir Aleksandrovich Fock was arrested in 1937. Kapitza found out about his arrest while he was in Leningrad. From there he immediately wrote to Stalin and to Mezhlauk. The letter to Stalin is short, explicit and scathing. It consists mainly of a list of numbered points, one of these points being a comparison of Fock's arrest to Einstein's exile from Nazi Germany.

Kapitza's main hope, of course, lay with Mezhlauk. In his letter to Mezhlauk Pyotr Leonidovich expressed his feelings simply and unambiguously, saying how Fock's arrest had shaken him and what an absurd blunder it was.

He insisted that Fock was a brilliant scientist whose attention was focussed only on science – in fact he was isolated from everyday life by his total deafness. Pyotr Leonidovich did not omit to mention that he had heard of the arrest of "a large number of theoretical scientists." "So many have been arrested," he wrote, "that in the faculty of mathematical physics (of Leningrad University – Author) there are none left to deliver certain of the courses... I trust that the investigations of the NKVD will show that the majority of these people have had nothing whatever to do with any kind of wrong-doing ... But what if they are proved guilty? That would be even worse... Most of them are young. It would mean that the Soviet state, despite having had twenty years to win over the scientists, has failed to do so ..."

Six days later Kapitza wrote another letter to Mezhlauk to express his gratitude – Fock had been released. However, before Fock was freed he was taken from his prison cell in Leningrad all the way to Moscow for a short meeting with Yezhov. The theme of this interview, which took place in Yezhov's office, was developed in a whole sequence of memorable slogans to the general effect that the enemies of the people were many and that constant watchfulness was the duty of all.

Yezhov ended the meeting with an apparently casual remark. Given the enormous amount of work that had to be done, it was, he said, inevitable that mistakes were sometimes made. However, his colleagues knew how to spot these mistakes and correct them. His – Fock's – case showed just how conscientious they were in this regard.

Landau's case, however, was not so simple. Kapitza did not receive any reply from Stalin. All his other efforts also proved fruitless. Years later it was revealed that Niels Bohr had also written to Stalin:

"... Over many years I have had the pleasure of corresponding with Professor Landau about scientific matters in which we are both deeply interested. However, to my great regret, I have received no replies to my last few letters and, as far as I know, none of the other foreign scientists who follow the progress of his work with the keenest interest have had any news of him, either. I have tried to get in touch with Professor Landau via the Soviet Academy of Sciences, of which I am honoured to be a member, but the reply I have received from the President of the Academy contains no information as to what has happened to Landau or his present whereabouts.

I am deeply concerned about this matter, especially as recent rumours suggest that Professor Landau has been arrested. I am still hopeful that these rumours are without foundation. However, if Professor Landau has indeed been arrested I am certain that this can only be the result of a grave misunderstanding, for I cannot imagine that Professor Landau, who has devoted all his energies to scientific research and whom I esteem so highly, could do anything which could justify his arrest.

In view of the fact that this is a matter of such importance, not only for Soviet Science but also for the whole international scientific community, I appeal to you most urgently to look into this matter and, if a misunderstanding has indeed arisen, to ensure that this outstandingly successful scientist can once more take up his research work, which is so vital for the progress of mankind." [46, p.7]

Landau had then been in prison for a year. At the end of the 1950s he wrote about that period:

"It was obvious to me that I could not last more than six months longer; the simple truth is that I was dying. Kapitza went to the Kremlin and said that if I were not released he would resign from the Institute. After that they let me go."

Kapitza did indeed go to the Kremlin; he was invited there to visit Beria, who had replaced Yezhov as head of the NKVD.* After this meeting Pyotr Leonidovich submitted an official request to Beria himself. It was written in the style that is expected in organisations such as that headed by Beria. In this letter Kapitza requested that Professor Lev Davidovich, then being held in custody, should be released on condition that he, Kapitza, would act as his guarantor. He, Kapitza, guaranteed that Landau would not involve himself in any kind of counter-revolutionary activity against the Soviet state either inside the Institute or outside its boundaries.

---

* Yezhov was shot, as was his predecessor. Stalin announced at the Politbureau: "A rotten man was Yezhov, so many outstanding people he ruined."

Before visiting Beria, Kapitza had also written to Molotov. The formal request addressed to Beria was written on 26 April 1939, the letter to Molotov on 6 April. It ended with a list of concisely formulated points:

> "... (1) Landau has been in prison for a year and the investigation is still not complete. The investigation is taking an abnormally long time.
>
> (2) I, as Director of the Institute in which he works, have not been informed of the charges brought against him.
>
> (3) The main thing is that, for no known reason, not only Soviet science but world science has been deprived of Landau's intellect.
>
> (4) Landau is in poor health and if his condition is needlessly made worse this will bring disgrace on us all.
>
> I therefore make the following requests:
>
> (a) that the NKVD be instructed to make all possible haste to bring the case to a conclusion;
>
> (b) if delay is inevitable, that Landau be allowed to apply his mind to some scientific work while he is detained at Butyrki. I am told that this is the usual procedure where engineers are concerned.
>
> P.L.Kapitza."

Landau was released. Yuri Borisovich, however, was by this time among the "engineers" – in fact, he was one of those very "engineers" who were "applying their minds to scientific work".

Professor Yu.B. Rumer's "Appeal" to the Presidium of the Supreme Soviet of the USSR, dated 12 February 1954:

> "I was arrested in April 1938 and in May 1940 sentenced in my absence by the Military Collegium of the Supreme Court to ten years' imprisonment. Throughout my ten-year sentence I worked for NKVD special section No.4 at a number of aircraft factories, heading a team which studied vibration fatigue.
>
> The conditions in which I was held were such as to permit me to pursue my scientific research work.
>
> At the end of the ten years, in 1948, I was sent to the town of Yeniseisk where I obtained a professorial post at the Teacher Training Institute. At the request of the late Academician S.I.Vavilov I was allowed to move to the city of Novosibirsk, where I now work in the local branch of the Academy of Sciences.
>
> Since 1948 I have published twenty-three articles on physics. In December 1952 I was summoned to Moscow to take part in a discussion of my work organ-

ised by the Academy of Sciences. The result was a recommendation that I should continue my research on elementary particles. In September 1953 the Ministry of Culture reinstated me as a professor and Doctor of Science and ruled that I should be regarded as having enjoyed this status without interruption since 1935. In view of the following facts:

(1) Most of the judgements against members of Special Section No.4 have been reversed;

(2) I am at the peak of my creative powers and that my work is now widely known among Soviet physicists and mathematicians – as two Heroes of Socialist Labour, Academicians I.Ye.Tamm and L.D. Landau, who know me well, can testify;

(3) The judgement made against me hampers my teaching activity in higher education, preventing me from passing on my knowledge to the younger generation;

I ask the Presidium of the Supreme Soviet to remove the judgement made against me and restore my full rights.

Yu.B.Rumer."

At the very end of 1954, just before New Year, Yuri Borisovich received a note from the Military Collegium of the Supreme Court of the USSR:

"The charge brought against Yuri Borisovich Rumer was reexamined by the Military Collegium of the Supreme Court of the USSR on 10 July 1954.

In the light of new evidence the sentence passed by the Military Collegium on 29 May 1940 is hereby rescinded and the case is officially closed" (seal with USSR emblem and the signature of the Lieutenant-General of the Judiciary).

Immediately after that interrogation during which Yuri Borisovich so eagerly agreed to work as an engineer his thoughts turned to science. Working as an engineer meant working at a desk, or not far from one – at the very least it would involve using pencil and paper. However, in Bolshevo prison, where he was sent straight from Butyrki, he did not get to the stage of actually starting work. The interrogations still continued. The main topic now was the scientific work that the prisoner had been doing before he was arrested and the purpose of the questioning was to assess the prisoner's suitability for some job or other, apparently a very important one. Right from the start of these interrogations Yuri Borisovich understood that the work in question lay in the field of aviation.

He did not, however, immediately guess that he was in Bolshevo. The place was a kind of sorting office: prisoners were being brought in and taken away all the time. Prisoners soon develop an instinct; they quickly learned to distinguish those who were brought from another prison and those who were in transit.

On one occasion, when Rumer was being marched back across the yard after a routine interrogation, he saw a small group of prisoners who had apparently only just arrived. "In transit" – thought Yuri Borisovich. The profile of one man standing next to the escort caught his eye. He realised immediately that he knew the man. It was not wise to show interest but Yuri Borisovich could not help taking another look. The man had turned away from Rumer, who now saw a haversack slung over his back with the three large letters "A.N.T" on it. Yuri Borisovich felt a stab of recognition – it was Tupolev! Neither that evening in the mess-hall, nor the next day, did Rumer see Tupolev again.

Soon Rumer himself was moved on from Bolshevo. The journey in the windowless van did not take long. The prisoner's keen hearing could make out that they had slipped around the outskirts of Moscow and then dipped back a little way into the noisy city. Then the van stopped. Would they be staying here for long? Yes, apparently this was their destination. They were at Tushino, where Yuri Borisovich was to begin his "scientific research". The inverted commas are meant only to indicate that this work consisted of two completely different parts – the overt and the covert.

The overt work was primary and fundamental, the work entrusted to him by the State. It was in a field of which he had no experience whatever. His background was in pure theoretical physics of a particularly quantum variety and, furthermore, he had come to physics "from mathematics". He now had to tackle problems in

applied mechanics and engineering. Nevertheless, Yuri Borisovich was
to achieve success in this field, too. He succeeded not only in his theoretical work on mechanics, on the theory relating to the oscillation of
complex systems, but also solved purely technical problems.

The State Archive in Novosibirsk preserves a report which originated in the Central Aero-Hydrodynamic Institute – CAHI. The report
records that "After the death of Academician A.I. Nekrasov in 1957,
A.N. Tupolev's Design and Development Bureau handed over to the
Zhukovsky Museum for safe keeping all of A.I. Nekrasov's papers
which he wrote during his time with the DDB. In the list of reports
dating from October 1938 to August 1943, which A.I. Nekrasov himself compiled, can be found the following:

> "... No. 46: A Theoretical Examination of the Wing in Non-stationary Airflow. In
> collaboration with Yu.B. Rumer
> ... No. 57: Flutter in Non-stationary Airflow. In collaboration with Yu.B. Rumer
> ... No. 60: The Application of the Theory of the Function of the Complex Variable
> in the Study of ... In collaboration with Yu. B. Rumer
> [47] etc.

Mstislav Vsevolodovich Keldysh began his work on flutter and
shimmy in aircraft wheels in Rumer's laboratory. When President of
the Academy of Sciences of the USSR many years later he recalled the
work done in those earlier days:

> "Rumer's articles on vibration in aircraft nose wheels are extremely valuable. His
> was the first work ever done in the Soviet Union on this problem. These papers
> present not only a theoretical analysis of vibration in the wheel but also a number
> of conclusions which are of direct practical use. Moreover, Yu.B. Rumer also
> conducted experimental studies of vibration using apparatus which he himself
> designed and built. These experimental investigations complemented and refined
> the theoretical work. His work, then, deserves to be held in high regard not only
> for its theoretical significance but also for the importance of its practical applica-
> tion." [47]

Yuri Borisovich was fascinated by this new field of research and
delighted to be able to check his theories immediately on his own
apparatus. He threw himself whole-heartedly into the work. Some-

times he remembered his father and thought that he was playing tribute to his father's wish all those years before – to see his son earning an honest living as an engineer. He also remembered Borka Venkov and his firm belief that the gods, in their own good time, always find a way of punishing those who betray mathematics. He worked long hours but the little time that remained, his own modest leisure time, Yuri Borisovich devoted to another project. This was to become the most important part of his life's work. It is called pentoptics.

Yuri Borisovich's scientific career was complicated, sometimes dramatic. We have seen how he returned to mathematics after a short, horrific period in the world of diplomacy. He returned to find Moscow mathematics in full bloom. Finding himself in the thick of mathematical developments in one of the best mathematical schools in the world he had every opportunity to achieve great things. However, he dropped mathematics for the fascination of the theory of relativity. He wrote a paper on a related theme and took it to Germany with only one aim in mind – to enter the new science and dedicate his life to it. He did not dream that he would meet Einstein – that would have been in the realms of fantasy – but fate decreed not only that he would meet Einstein but he would even be recommended to him as "the ideal assistant". However, this undreamed-of collaboration with the greatest of physicists did not take place. Why not?

"When for the second time, a few months after the first meeting and on Born's direct instructions, I travelled to see Einstein," wrote Yuri Borisovich, "I was already a fanatical adept of the quantum faith and nothing else existed for me. This time our conversation lasted about an hour. Einstein explained in detail his work on absolute parallelism (one of the versions of the "unified field theory" – Author). My attitude to these ideas was then much the same as that of other young "quantum" physicists – i.e. I did not believe that a "unified field theory" could be constructed. However, I did not have to admit this openly to Einstein; he evidently realised as much without a word from me. What I, on the other hand, did not realise was that my indifference to the ideas which so gripped Einstein drew a line under that section of my biography which was connected with him." [18, p.111]

The final answer probably lies in those words: for "the fanatical adept of the quantum faith" nothing existed outside quantum theory. Being easily roused to new enthusiasms, so exceptionally able in so many different ways, he was able to lightly cast aside something which had seemed to be the main thing in his life only the day before. He was able to cast it aside at the very moment when the dream became reality, when he was on the crest of the wave. His love was universal, without favourites: physics, mathematics, chemistry, literature, girls, languages. He did not dole out shares of his energy or calculate his own advantage – he simply threw himself headlong into each of his passions. For this the gods would more than once find opportunity to punish him.

Returning for a moment to the Göttingen period it must be said that at the height of the "quantum craze" Yuri Borisovich did glance briefly at the "unified field theory". He dashed off one paper but did not even try to get it published. However, it was this paper which prompted Max Born to write to Einstein:

"Dear Einstein!

By the same post I am sending you Rumer's new paper in which, I think, he has made a real step forwards in the direction in which he has been heading for several years now. I realise, of course, that your thoughts are focussed on entirely different things but perhaps you will find time just to look through Rumer's paper.

   I think that his assertion is correct – i.e. if Riemann's space is accepted then the inescapable consequence is that other concessions have to be made as regards the tensor of matter and, with this necessity established, he goes on to outline a novel field theory of matter.

   Only one question now remains – i.e. is it a good idea to continue in this direction and shape this theory properly or is it better to move on to a completely new kind of geometry, which is what you have attempted? I am no judge in this matter but I think that both paths should be followed.

   Cordial greetings from me and from my wife.

Yours, Max Born" [2, p.21]

Nothing is known of Einstein's assessment of Rumer's paper. This letter, completely devoted to Rumer, turned out to be the last that Max Born wrote to Einstein from Göttingen. Born wrote the following comment about this letter:

> "My letter of 6 October 1931 and Einstein's next letter to me were separated by a period of about eighteen months which was crammed with so many dramatic events that scientific matters were pushed into the background. This was the period, already mentioned, when I served as Dean. There were several elections to the Reichstag, the result of which was an increase in the number of Nazi deputies and the strengthening of Hitler's influence. Mobs of Brownshirts terrorised the country. Then came the Nazi takeover. On one occasion at the end of April 1933 I found my name in a newspaper. I was listed among those who were disqualified from state appointments by the "New Laws". [2, p.22]

By that time Yuri Borisovich Rumer was already at home in the midst of the happy upheaval that had so transformed Moscow. The paper which Born had considered to be a "real step forward" in the direction to which Einstein was committed now seemed like an insignificant episode in the distant past.

And then, of course, came the time when Yuri Borisovich was compelled to tackle a completely different but very important set of problems. As Rumer worked alongside Tupolev, Korolyov and Stechkin – and as he took his place in the prison string orchestra which gathered in their brief moments of leisure and for which they made with their own hands violins and violas from scraps of aircraft-quality plywood – Einstein began to seem to him like a character in an ancient myth. And yet this was the time when Yuri Borisovich turned back to his old ideas and became obsessed with them and that distant episode gave him the impulse to start on an enormous project which was to last one and a half decades.

The central idea of the project went back to Kaluza and Klein and to the thinking of Einstein and Bergmann concerning the use of the concept of five-dimensional space in order to bring together both electromagnetic fields and gravitation in a unified description.

Kaluza's paper, in which he constructed a single theory of electromagnetism and gravitation in five-dimensional Riemann space, was published in 1921. The trajectory of a charged particle is interpreted as a geodesic line in this space. The most surprising discovery was that the additional equations which described the curvature of space in the fifth dimension coincided exactly with Maxwell's equations. The fifth coordinate is cyclical. We (convinced of the four-dimensional nature of our world – three spatial dimensions plus time) cannot perceive it because it is twisted into a ring of very small radius. The electron is the messenger which announces its existence to us. Einstein's letters to Kaluza have been preserved. (In 1919, before publishing his paper, Kaluza sent it to Einstein). In these letters Einstein, though excited by the elegance of Kaluza's ideas, expressed certain doubts. It took for Einstein two years to accept Kaluza's ideas and publish the paper.

In 1926, after the discovery of quantum mechanics, Oskar Klein, borrowing Kaluza's idea, further developed the five-dimensional theory of electromagnetism and gravitation. Einstein, then completely absorbed in his search for a unified field theory, was trying out a wide variety of approaches. One of his avenues of enquiry involved rejecting Riemann's dimensions and instead developing a more general geometry.

Towards the end of the 1930s Einstein came back to Riemann's five-dimensional space. In 1938 he published a paper which had much in common with the work of Kaluza and Klein. Formally, the theory of Einstein and Bergmann included a description of both electromagnetic and of gravitational fields but Einstein himself considered it to be artificial since it left open the question as to the physical meaning of the fifth coordinate.

Einstein devoted the last thirty-five years of his life to his attempt to create a unified field theory – the theory which, he intended, would explain both electromagnetic and gravitational effects in terms of the same basic principles. He did not succeed. Hopes for such a theory remain only hopes to this day, although the work now proceeds on a very different level.

It was with this far-from-simple topic that Yuri Borisovich, despite his complete isolation, now chose to wrestle. He discovered that to the fifth coordinate can be ascribed the physical meaning of action and to its period the numerical value of Planck's constant. He wrote:

> "This leads to a far-reaching synthesis of the geometrical concepts that are implicit in the general theory of relativity with the ideas of quantum theory" and "the problem for relativistic mechanics (classical and quantum) of the motion of a point in a gravitational or electromagnetic field is analogous to the problem in optics of the diffraction of rays in five-dimensional Riemann space" [42, p. 9].

And so Yuri Borisovich developed his "Pentoptics". He was sure that he had made an important discovery.

For ten years he nourished himself with this work, checking and re-checking it, each time convincing himself that the theory was correct and free of contradictions. He wrote his notes on sheets of thin photo-sensitive paper which he sewed together in the middle with a needle and thread to form a large yellowish-pink note book. He built up a stack of such books containing in total more than three hundred pages of notes. He would cover a pink sheet before going to bed and fall asleep with his mind full of the material which would fill the next sheet the following evening. Even his dreams, when he had them, were coloured yellowish-pink.

Even poems were yellowish-pink. No matter where the Special Design and Development Bureau was moved to, it always possessed a magnificent technical library. Literary works, however, were in shorter supply and poetry was particularly scarce. On their rare free evenings, therefore, they would gather and recite poems from memory, then write them down on the yellowish-pink sheets and sew them together in anthologies. Yuri Borisovich's family still preserves "Anna Snegina" and three chapters of "Pentoptics" from the yellowish-pink library. The rest were lost. Yuri Borisovich had to re-write "Pentoptics" during his exile. He did this quickly, without any self-pity or bitterness. He wrote from exile to Tanya Martynova:

"My dear Tanya!

> I am constantly touched by your attentiveness towards me and appreciate it very much. I have to overcome a kind of psychological block in order to force myself to sit down and write. I really do not know why. The main thing that sustains me is the deep conviction that I have made a major scientific discovery and fully justified the hopes which people had for me in my youth ... I am doing everything that it is in my power to do and I hope to live to see my work recognised ... Circumstances force us to live alone, without seeing anyone, completely left to our own devices. My job at the Teacher Training Institute keeps me alive. I feel a kind of infinite strength at every level of my being: it is dreadful to think how many graduate students I could now keep occupied with challenging tasks. The only two teachers of mathematics here have already sent off three papers written under my supervision – two have been published and the third is being printed. That is all I have to say about myself ..."

The person to whom this letter was addressed, Tanya Martynova, was a close friend of Yuri Borisovich. He had become acquainted with her parents after his return from Germany in 1932 and became a firm friend of the family. Tanya was then only a girl. By the time Yuri Borisovich was arrested she was a student in the Geological Faculty of Moscow University.

The first meeting held at the University after the May holiday in 1938 was dedicated exclusively to Rumer. These meetings at which the academic staff gathered to condemn former colleagues, now enemies of the people, were usually held when a sizeable batch of such saboteurs had been rounded up; otherwise the gatherings would have been rather too frequent. The meetings did not take long – the denouncers denounced and the rest kept quiet. If, unexpectedly, anyone wanted to say anything he or she would be given the floor without demur.

The meeting at which Rumer was denounced did not differ particularly from all other such meetings: the denouncers denounced and the rest kept quiet. Then at the end of the meeting a slim young girl student asked permission to speak. Permission was granted with calm indifference. The girl was Tanya Martynova. "Comrades," said Tanya, "I swear to you that everything that has been said here about Yuri Borisovich Rumer is untrue! Let's all stop and think about what we are doing..." She spoke quietly, almost in a whisper, which increased the tension in the hall yet further. She spoke in such a way that no one dared to interrupt her and it was only when she broke down in tears, frightened by the dreadful silence and the faces all stiff with horror, that the meeting was brought hurriedly to a close. After this meeting all Tanya's university friends deserted her. Some expected her to be arrested, others apparently regarded her as an "agent provocateur". However, Tanya was left alone, thank God.

Returning to the letter addressed to Tanya Martynova, let us note one important part of Yuri Borisovich's character which is reflected therein – his unswerving desire to find and teach his own students. Whenever he met young people he would unconsciously assess them as potential successors. Although it lay dormant for a while after his arrest this instinct soon reasserted itself even at Tushino.

One evening when almost everyone was already in the dormitory, each absorbed in his own preferred activity, a very young man was brought in. He was shown the place that had been prepared for him. In one hand the young man held a book by Pontryagin, in the other an empty-looking haversack, from which he retrieved a piece of bread. Holding the bread on a half-outstretched palm, he started to look about him. This sight could only be fully understood by someone who knew how it felt to be brought back from the camps to a world where people "had it easy".

Rumer had never been in a camp but he was touched to the heart and was the first to go up and speak to the youth. It was immediately revealed that Yuri Borisovich was partly responsible for his releas from the camp. The young man's name was Kolya Zheltukhin.

"I was arrested in 1937," wrote Nikolai Alekseyevich Zheltukhin, "It was a long-winded business – first the interrogations, then the trial, then the waiting for the decision of the Court of Appeal.

In 1939 my appeal was rejected and I was sent to a camp in Kotlas, not in Kotlas itself but to a logging gang whose job was to float timber down the river Sukhona and its tributaries.

The area where we worked was fenced off. Our job was to push the logs with our boathooks along the smaller channels towards the machine which tied them in bundles ready for the big river. We lived in a barge on the river. The bank was guarded and there was water all around us, icy-cold water. I realised then that a human being can endure much more than his reason might suggest.

There I wrote a description of an invention which would improve engine ignition systems. It was intended mainly for aero engines but was also applicable to automobile engines. This document of mine went via the prison administration to Moscow where it was directed to Stechkin.

Someone in Moscow looked through my scribbled notes – and they literally were scribbled on sheets from school exercise books with freehand sketches instead of proper drawings. It must have been hard to make sense of this scrawl. Nevertheless, it would have been clear that the person who wrote it knew something about the subject, whether he was a professional or not. They came to a swift decision – quite a risky thing to do in those days – and the prison authorities called me to Moscow. The document stating the decision was signed by Professor Stechkin and Professor Rumer.

When I got to Moscow the bureaucrats immediately became suspicious – I was too young, only twenty-three years old. All the same, they sent me to Tushino.

It did not take the people at Tushino long to realise that I was not a professional engineer. However, I was an experienced draughtsman. In my student days I had earned extra money by working in a factory design office and I had developed a steady hand on the drawing board. They kept me on at Tushino and gave me the job of producing a general arrangement drawing of one of the engines.

Two types of engine were made at Tushino. One was being developed by Dobrotvorsky, the carburettor expert, and the other was the work of the famous designer Charomsky, under whom Stechkin and Rumer were working. They had all come to Tushino from

Bolshevo. The Bolshevo period was over by the time I arrived but they were still always talking about it. As far as I could make out, Bolshevo was a kind of staging depot where all the prisoners were held while the authorities decided who would do what. Then they were sent out to particular factories and design offices behind the bars again where their real work began.

I arrived at Tushino in July or August 1939. Yuri Borisovich Rumer immediately took me under his wing. He was busy with calculations relating to Charomsky's diesel engine. This engine had no fewer than four crankshafts. Because of this unusual complexity the engine was prone to all kinds of oscillatory processes. Rumer was studying torsional vibration in the crankshafts.

He was doing other jobs, too. He worked in collaboration with Stechkin, in particular, but I do not know what became of their work – that is, I do not know who is credited with that project. There is no doubt, however, that the results were successfully applied in practice.

Yuri Borisovich was very keen that I should be involved in calculations but in Dobrotvorsky's office, where we were working on a conventional multi-cylinder carburettor-aspirated engine, there was no such requirement and I stayed with general arrangement drawings. However, Yuri Borisovich constantly discussed his work with me and later, when I did some of the calculations for a new engine at Kazan, I used certain of his approaches and found that they worked.

By that time, though, we had lost touch with each other. I could not have imagined then that we were destined eventually to be neighbours in the same little town for more than twenty years. And although we were together at Tushino for no more than six months we always looked back to that period.

Without exaggeration, Tushino was my salvation. I had been arrested when I was a third-year student. The charge against me read: "Anti-Soviet agitation, Article 58, Section 1". They gave me eight years, plus five years loss of rights. Before they got round to pronouncing sentence, though, they held me in prison at Voronezh, then at Boguchar. When the sentence came through I was sent to the camp. And then, after all that, I ended up at Tushino.

At Tushino there was a clean yard and clean wooden buildings. There was a well-kept single-storey building which housed the dormitories as well as the offices where the designers worked on drawings and calculations. Nearby was the factory where our engines were produced – but I never went there. There was a big, sunny canteen, with one big, round table that was covered with a table cloth, or maybe oilcloth, but always clean. They served good food three or four times a day – breakfast, lunch and supper and between breakfast and lunch there was a tea break.

The whole country was well-off as regards food at that time and this was reflected in our canteen. You can judge how well we were fed there by the fact that I recovered from tuberculosis while I was at Tushino. Just the excellent diet cured me. I had arrived from the camp suffering from tuberculosis. I did not know that I had this tubercular process affecting my lungs. I just knew I had a constant cough and was "going home" as they said in the camps. I got thinner and thinner and thinner. Then when I arrived at the design bureau and started eating at the canteen – where they served meat for lunch and supper, butter, yoghurt etc. – I quickly got better. Five years later and on several occasions after that I was asked at medical examinations when the process in my lungs had stopped. I realised that it must have been at Tushino.

There was good "spiritual food", too, or, more precisely, technical spiritual food. There was a big library at the factory and although we were not allowed on the factory site there was an entrance to the library made specially for us. They brought fiction and poetry from Butyrki – not much, though, and the books were changed only every two or three months. The library at Butyrki, by the way, was excellent; it was constantly being replenished with books seized during searches and arrests.

In late 1939, or perhaps at the beginning of 1940, Yuri Borisovich was transferred to Tupolev's staff in Moscow. He wrote to the prison authorities requesting a move from the engine design centre at Tushino to Tupolev's airframe design bureau. Evidently there was a supporting request from their end, too, so the application was successful.

Only Yuri Borisovich was transferred. I tried to move to the airframe DB, too – mainly because I wanted to be with Yuri Borisovich; he was, after all, a professor of physics, an interesting and knowledgeable man and I was drawn to him – but they would not let me go.

We were at Tushino until the summer or autumn of 1940. Then it was decided that our engine should be put into production at a factory in Kazan so Dobrotvorsky's team, myself included, moved there.

Glushko was a member of our team. While we were still at Tushino I was present when Glushko suggested to those in charge of us that a separate DB should be set up to work on rocket engines. This suggestion was considered but it was turned down. However, Glushko would not give up even though it was his interest in rocketry that had got him into trouble. It had got Tuchachevsky into trouble before him, too, and then Kleimenov and Langemak were imprisoned and shot for the same reason. Then Korolyov was put in prison. Nevertheless, Glushko got his way in the end. While we were in Kazan it was decided that Glushko was to be put in charge of his own separate DB.

As I had some ability in mathematics and a good educational background I was transferred to Glushko's new DB to work on the calculations for the rocket motor. Calculations for a completely new type of engine are a real challenge. I may not have finished my degree course but what a university education I had! There were teachers all around me – Stechkin, Rumer, Glushko, Professor Pazukhin, then Korolyov!

We worked twelve hours a day. Every one of these men was more than just a brilliant scientist – each one was also totally committed to the job in hand. And these were my teachers. I became a sort of corresponding member of the Academy of Sciences without even possessing a degree certificate!

Glushko's DB was set up in 1942. Korolyov came on a working visit. He travelled from Omsk prison under guard. This was before the rocket motor was given the go-ahead. Korolyov was then on the staff of Tupolev's DB, where Rumer worked. Korolyov and Glushko talked and wrote notes for some time. Apparently they had joined forces to push for the rocket DB. I say "apparently" because nothing was said openly about this. The war was on. We were working very hard. Shortly afterwards Korolyov came to Kazan to work with us, but still as a prisoner.

In less than two years the Glushko DB made great progress with the LFRM (liquid-fuelled rocket motor). Tests had demonstrated its viability and its potential. Then, in 1944, a large group of prisoners, several dozen people, were set free. Among them were Glushko and Korolyov. I myself was not on this particular list. Meanwhile Moscow decided that the rocket propulsion programme should be expanded. Several new factories and design bureaus were set up. They were organised in two divisions with Korolyov in charge of one and Glushko in charge of the other. I was put with Glushko and I worked under him until 1959.

In June 1945 my sentence came to an end. They usually released people on the exact anniversary but in my case it was three days later – that is, I served eight years plus three days. From then on I was a free employee in the DB. However, I still had five years' loss of rights in front of me and according to the strict letter of the law I was allowed to live only in specified areas. However, Glushko kept me with him. I worked with him for seventeen years altogether.

In ten years our DBs – Korolyov's and Glushko's – had moved so far that the larger sections of the Academy could not keep up with us. This was not just a technical achievement. It was a massive scientific research programme. In an unexplored field there is no difference between serious research and the work of a design bureau – except that in the DB the work is so purposeful and the pressure so intense that it is difficult to describe to anyone who has never experienced it.

All this talk about people working because they were afraid – it is all completely untrue! People do not work like that out of fear – they did it because their hearts were in it. In addition to that, you have to consider the unity of purpose, the total or almost total absence of individual egoism and the staggering expertise that these men brought to the job.

No bitterness was ever expressed openly – but, of course, it is hard to judge what goes on inside the human heart. We never discussed grievances. Each had his own thoughts and it was considered improper to talk about them, especially as in many cases it would have involved talking about the "persons" who had had people put away or who had facilitated their arrest. Such persons acted as they did either out of spite or fear or because they had been forced to contribute to a string of false accusations. We did not judge them; in fact, we hardly ever mentioned them. We all shared the same unshakeable faith: that one day we would be proved innocent.

The only time I ever went to Tupolev's DB on Radio Street – I forget why I was sent there – I ran into Yuri Borisovich. We were delighted to see each other and he treated me to an enormously enthusiastic account of his work in aerodynamics – the mathematics of wing behaviour and of parasitical oscillation in aircraft nose wheels. He introduced me to Krutkov and Bartini. Bartini sat withdrawn into himself at his drawing board looking a little like an exotic bird in a cage. We others, however – Rumer, Dobrotvorsky, Krutkov etc. – were full of optimism. We had an interesting job to do and we lived constantly in the hope that we would soon be released. Had it not been for the Finnish War we would probably have been set free much sooner.

Yuri Borisovich and I were not to meet again for twenty years after our encounter on Radio Street. In 1959 Khristianovich invited me to visit him in the institute he had just opened at Akademgorodok near Novosibirsk. I accepted his invitation. Yuri Borisovich was director of IRE, the Institute of Radiophysics and Electronics in Novosibirsk, and was then in the process of moving his institute to Akademgorodok. After that I stayed on permanently in Akademgorodok and so did Yuri Borisovich. He died in 1984. I know for a fact that everybody loved him."

Tupolev's DB (and then prison. Today it is his son's DB) on Radio street in Downtown Moscow

Chapter 14

Earning a Living as an Engineer

On Radio Street in Moscow stands a huge building that was constructed in the early 1930s. The back of the building looks out on to the River Yauza and on to a riverside thoroughfare which is today named Tupolev Embankment after the aircraft designer who was the building's first occupant.

In the 1920s and 1930s CAHI, the Central Aero-Hydrodynamics Institute, followed a rather abstract and academic path in aeronautics. Tupolev at first worked within CAHI but later his design bureau was detached from it and became an independent organisation. The task entrusted to Tupolev's independent bureau was the development of experimental aeroplanes. Aircraft were soon being built in his own factory and tested by his own pilots.

The huge new building on Radio Street was provided to house Tupolev's DSEAE – the Design Section of the Experimental Aircraft Establishment. The building shot up and was very quickly fitted out, each floor being designed and equipped to serve a particular purpose. All the work was done to the highest possible standard. All eight floors sparkled with up-to-the-minute features and large windows let in plenty of light.

Although it was headed by a man whose habitual facial expression resembled that of a kindly story-teller, the organisation was far from being cosy or restful. Here the concept of a routine "working day" was alien. The staff, which included men like Archangelsky, Myasishchev, Petlyakov and Sukhoi, might work for 24 hours at a stretch or even for weeks on end.

Tupolev was arrested at his desk in the DSEAE. Three men entered the chief designer's office at about 11 p.m. one evening and at some time after midnight led him out into the street. They are not likely to have suspected that one day that street would bear their distinguished prisoner's name. They may have lived long enough to fly in his aeroplanes. Those who gave the order for Tupolev's arrest and those who interrogated him, as well as those who fixed bars at all the windows on all eight floors of the DSEAE, may have lived to fly in various Tupolev types. Then again, maybe they did not.

It took the authorities less than a year to complete the conversion of the DSEAE into a research centre where all the scientists were to be convicts. In that time all the windows were barred on the inside, the eighth floor became Prison Administration Office No. 29, prisoners' dormitories were constructed on the seventh floor, the prisoners' canteen was installed on the sixth floor, the 'monkey house' – the prisoner's exercise yard – was built on the roof, all free employees were brought back and a new boss moved into Tupolev's office on the third floor – a certain Colonel Kutepov of the NKVD, a former electrician.

Prisoners were sent here from various places, from Bolshevo and Tushino, from prisons and camps. Their attire also varied enormously. Some arrived in blue flying overalls, some in prison uniform, one even wore tattered felt boots and a blue and white striped mattress cover instead of a coat. Later they were all issued with blue flying suits and each was assigned a bed with a flannel blanket.

In the canteen there were neat little tables, each big enough for four diners. The tables were laid with white tablecloths – and even flowers appeared on them occasionally. Newcomers, accustomed to brutally cramped cells, primitive staging posts and camps in the taiga, could not believe their eyes when they saw all this luxury. In addition to the material comforts the majority of the prisoners – of whom there were over 150 in total – were former colleagues from CAHI and many were already friends. After being torn away from their work and subjected to dreadful experiences they were now suddenly together again and could get on with the work that was their life.

The technical staff – the laboratory assistants, the draughtsmen, the people who did the painstaking, detailed work which has to be done in any big design office – were practically all people who had worked in the DSEAE before the transformation. Now, however, they were called "free employees" to distinguish them from the prisoners.

Apart from the leading aircraft designers, this new prison gathered in physicists, mathematicians, and any specialists from other fields who might be useful. One distinguished theoretical physicist brought to Radio Street was Yuri Aleksandrovich Krutkov, who had taken part in Ehrenfest's Petrograd seminar. Also among the company were the mathematician Pyotr Aleksandrovich Walter, Karl Szilard the Hungarian mathematician, Aleksandr Ivanovich Nekrasov the expert on mechanics and hydrodynamics, Roberto Bartini the aircraft designer – who considered himself just as much a physicist as a mathematician – and Yuri Borisovich Rumer. They were all close friends. Yuri Borisovich was the youngest among them.

Aleksandr Ivanovich Nekrasov had been deputy to the head of CAHI at the time of his arrest. He had been a corresponding member of the Academy of Sciences since 1932 but was accused of being an American spy. He was arrested immediately on his return from a visit to America where he had been injured by a car in a street accident and had spent some time in hospital. Yuri Borisovich started working with him in the DSEAE and in 1945 they published a joint article (State Technical Theoretical Press – Gostekhteoretizdat): "A Theoretical Study of the Wing in Turbulent Airflow". Nekrasov received the Stalin Prize for this work in 1952. Rumer was not mentioned in the citation as he was still deprived of normal rights at that time. Aleksandr Ivanovich wrote to Rumer about the prize and asked him to accept part of the money which, he emphasised, was "due to you" – but Rumer refused.

Krutkov and Pyotr Aleksandrovich Walter were admitted as corresponding members in 1933 ("born in the same litter", to use Krutkov's expression). Krutkov was one of the few who changed while in captivity. From time to time he would be his old self – witty and light-hearted, even devil-may-care – but not for long. He worked hard and long, just like everyone else, but his deeply-set eyes under the enormous, heavy forehead seemed to gaze far into the distance. Of all the prisoners, he was least often to be seen walking in the "monkey house". "I never did like Moscow," he said, "and when you are looking at it through bars it's an even more sickening place to walk around." How different were Rumer's feelings towards Moscow! He always loved to get up on to the roof and "walk around Moscow", as Krutkov put it. However, he tried to keep away from the western side of the roof. That was the side from which he could see the whole of the Maroseika, the Church of SS. Cosmas and Damian and the land that had belonged to the Lutheran church, the land that lay opposite the Yegorovs' apartment house, where he had been born and raised.

From the Radio Street days onward Yuri Borisovich spent practically the whole of his prison term with Walter. It was while they were still in prison together that Walter died. This was in 1947 at the last of the factories where they worked together. Pyotr Aleksandrovich died of sheer exhaustion. It was a factory like any other – the only difference was that the workers were all on the edge of starvation.

Special Section No. 4 of the NKVD – in prisoners' language the "golden cage" or "shady outfit", which Beria set up in the former DSEAE and supervised personally, consisted of three design bureaux under three chief designers – Tupolev, Myasishchev and Petlyakov. Later a fourth DB was added – that of Tomashevich. This is where Tupolev's dive-bomber, Petlyakov's famous "100" (which became the Pe-2) and Myasishchev's long-range bomber were all designed.

Even in the 1920s it had become obvious that the design bureau needed more space. Tupolev had persuaded the Moscow City Soviet to

grant him a healthy chunk of territory and then had started the construction of a two-storey building which was, for its day, of grandiose
proportions. This was CAHI's experimental factory. He then built the
structure which included the wind tunnels and the tower upon which
the wind laboratory was placed. Tupolev's programme was simple:

> "The country needs aeroplanes as much as it needs black bread. It is no good
> offering flans, cakes and tarts if you have not got the ingredients with which to
> make them. Consequently we have to
> (a) work out an aviation policy based on the use of machines which can actually be
> built;
> (b) make use of the expertise and the productive facilities which we actually have to
> build planes that are suited to mass production ... it is vital to develop in
> every way possible the technology of experimental aircraft construction.
>
> In the Soviet context midget design bureaux, even if they are led by talented
> designers, will not be able to achieve much. They will not have the strength to
> push their way through all the bureaucratic obstacles. We need large-scale or-
> ganisations like DSEAE. Two, or perhaps at most three, such units would suf-
> fice." [49, NS, p.41]

And so Tupolev built the 8-storey building for the DSEAE. His
next project was a new factory. This was the EDF (Experimental Designs
Factory). With characteristic vigour he proceeded to lay the foundations
of more new buildings until he had annexed all the unused land adjoining that which had been granted to him by the Moscow Soviet. But even
Tupolev, with all his fantastic intuition, could not foresee that this immense complex, in the building of which he had many times put his own
head on the line, would turn out to be an ideal base for a special prison.

Leonid Lvovich Kerber, a fellow-inmate of Tupolev's in the "outfit" and later his deputy, writes:

> "I came to DB-29 from the camps around Archangel where I had been felling
> timber for about eight months. Before that I had been in prison for about a year
> – first at Butyrki, then a short trip to Lefortovo to sample its horrors. They came
> to fetch just me alone from the camp. I was told to get ready to move in the usual
> fashion: "Get your things!" I travelled by rail in an overcrowded "Stolypin" van
> – a windowless freight car. It was in this van – "in my compartment" – that I met

Yuri Borisovich Rumer and Yuri Aleksandrovich Krutkov. The train stopped at Vologda and two more prisoners were pushed in. The first to get in was a tall man with a handsome, manly profile. He was wearing spectacles from which one of the lenses was missing. He was wearing a strange sort of fur creation on his head and a sleeveless body-warmer. Over the body-warmer he wore a mattress cover.

Behind him stood another, shorter man who also had a striking sort of face. This turned out to be Yuri Aleksandrovich Krutkov. This was how Rumer and Krutkov looked when they were thrust into the van.

For a long time I tried to recall where and when I had first seen Yuri Borisovich and could not remember. We lived then in another dimension, in an unreal world. The only realities were the bread ration and the roll-up cigarettes which we made for ourselves. Trying to remember was like developing a film or plate in a dark room. You shake the developer dish and gradually you start to make out an image. It was just like that in this case. The memory of that scene gradually developed in my consciousness.

We talked for a long time. When you are living in conditions like those it is a real joy to meet someone from the same background as yourself. The event alone is a joy – but then you discover that there is an endless stream of things to talk about. We talked all the way to Moscow, all the time completely ignorant as to where we were going or why we were being taken there. All I knew was that they had been brought from camps in the Mariinsk-Kannsk area. At the Butyrki in Moscow we were split up. It was a prison rule that people who met in transit should never be allowed to stay together at their destination. We were taken off to different cells. Later, though – it must have been March or April 1939 – we met up again in DB-29 on Radio Street. We greeted each other like long-lost kinsfolk, embracing and kissing."

A word-by-word transcription of L.L. Kerber's conversation with the author would have been a disturbing departure from the style of the book but it would at least have enabled the reader to realise that Yuri Borisovich Rumer could not possibly have been one of the two prisoners who entered the "compartment" on the occasion described. Rumer never spent any time in any camp. As we know, he went from Butyrki straight to Bolshevo and by the time of the episode on the train from Archangelsk he was already at the Tushino aero-engine factory working in the design bureau of his fellow-prisoner Stechkin.

In 1988 the journal "Inventor and Rationaliser" (No. 3-9) published Leonid Lvovich Kerber's recollections of his time at DB-29. The reader will find there several more stories about Rumer and Krutkov. In actual fact Yuri Borisovich could not have been involved in any of these events, either. However, this does not matter very much. When the misunderstandings were cleared up Leonid Lvovich's reminiscences were already in print. No doubt the mistakes could have been corrected but Leonid Lvovich decided not to change anything. He and I agreed that the main thing was to remember that these things happened – even if they did happen to other people.

The one really annoying error lies in the description of Yuri Borisovich as a "dyed-in-the-wool misogynist". If he had seen himself thus described he would have been most upset. It is also necessary to set the record straight on two other points: Krutkov was not 70 years old as is claimed – he was not even 50; finally, neither Rumer nor Krutkov nor Karl Szilard ever worked in any atomic research "outfit" – simply because there was no such thing. Since it was agreed that nothing should be changed this last rather serious inaccuracy was also retained. Plenty of other secret organisations existed, of course, but not this one.

On Radio Street, then, there were three design bureaux: DB-100 (Petlyakov) DB-102 (Myasishchev) and DB-103 (Tupolev). The aircraft that were produced by these teams had not only their own names (for example Tupolev's Tu-58, so called simply because it was his 58th design but also appropriately named – as he himself pointed out – since it was created under the auspices of Article 58!) but also affectionate nicknames bestowed on them by the prisoners (we shall meet "Vera" later on, for example). They also had official model designations: 100, 102, 103.

The scientific section was at first shared by all three bureaux but since it was so huge it also was divided into separate teams, each concerned with a different aspect of aeronautical theory. The "theoreticians" who worked in this section included Nekrasov, Walter, Rumer,

Krutkov, Szilard, Ozerov, Sokolov and Sterlin. Some of their colleagues have still not been officially rehabilitated to this day.

"The life of the 'zeks' (prisoners) gradually assumed a settled pattern," wrote Leonid Lvovich. "As long as we were actually at work on the 103 we were free of those inner emotions of which we could no longer remain unaware once we were back in the dormitory, in our own world to which the "bums" were denied entry. This settled pattern was disturbed for the first time when Petlyakov and part of his DB staff were set free. This followed the successful trials of his "100". This decision, though of course welcomed, at the same time added to the psychological burdens of those who were not going to be released. There is no need to go into the negative side here as it has been covered fully in my reminiscences. I will just say that hopes were raised and we ourselves voluntarily lengthened our working day just to get the aircraft completed more quickly."

In his reminiscences Leonid Lvovich wrote:

> "Taking shape on the building jigs were the aggressive jaws (the long bomb-aimer's window under the nose) and snout of the 103. The rear section of the fuselage was surprisingly slim and graceful, as were the tailplane and twin fins. The wing centre-section formed a massive torsion-box. The outer wing panels were removable. The plane had long twin engine nacelles from which emerged the slender undercarriage.
>
> It is no exaggeration to say that to the eye of the engineer the prototype 103 represented the height of elegance. Striving for the maximum possible speed, Tupolev had cut down frontal area to the absolute minimum and there is no doubt that the slim fuselage contributed to the 103's good looks.
>
> There was not one spare inch in the crew positions but nevertheless, the Old Man (Tupolev) demanded elegant design even there. There is a story about this. One evening two zeks went into the mock-up shop. Usually it would have been empty at that time, and they were startled to hear a crack from the cockpit followed by the squeak of nails being extracted. Then a panel was torn from its place and thrown out. It traced a parabola through the air and fell on the floor. The prisoners stepped up on to the jigs surrounding the mock-up and discovered that the Chief was at work on improvements to the 103. They could hear him muttering to himself: "What a mess! Each one of them has stuck his own bits and pieces in here. Call this a

cockpit? More like an ...! A man is going to do a job in here – maybe die while he's doing it – and this lot, instead of making his work easier, have lumbered him with God knows what!" Heralded by these words and the squeak of more nails, another plate or panel followed a curving trajectory from the cockpit to the floor.

During one of these modification operations the lion's share of the redesign work fell to our weapons specialist A.V. Nadashkevich. A whole lot of things had to be worked out again from scratch and Nadashkevich went to the chief to complain. Always ready to see the humour in any situation, the inmates immediately composed a song in which Tupolev, determined to make a few improvements, sings:

"Nadashkevich, my old mate,
What you've drawn is simply great!
There's just one tiny mod to make;
It will be a piece of cake.
Not a long job, very quick,
One rivet there will do the trick.
Of course, the longerons will be longer,
So the skin needs to be stronger
And the engines move to here
To make sure the props are clear ..."

The long list of modifications ends thus:

"Apart from that, my dear old mate,
What you've drawn is simply great!"

When he heard this song Tupolev laughed out loud and rashly promised: "Damn the lot of you, I won't make any more changes!" – but he soon returned to his old ways. In such cases, if anyone said to him: "But Andrei Nikolayevich, what about the plan, the deadlines, the blueprints?" he would interrupt sharply: "Does it say in the plan that we have to make a pig's ear of it?" [49, No. 7, p. 40]

That is how they built their "103". If anyone in the team, whether designer or theoretician, could not sleep he would go down once more to the mock-up shop, or to the rooms where the calculations were done to look once again at the formulae. Everyone had the same aim – to get the plane finished. Nobody had the slightest doubt that she would fly – and fly superbly. They did not know what difficulties lay ahead for the 103.

Leonid Lvovich writes about the completion of the prototype:

"Finally everything on the plane had been checked, adjusted and tested. Only the engine test remained. The next day the outer wing panels were unbolted and the 103 was rolled out through the doors of the assembly shop into the yard. Hundreds of people stood at open windows to watch their baby take her first steps. The 103 shuddered like a thoroughbred racehorse before the off. Dressed in a warm coat and smiling happily, Andrei Nikolayevich shook hands with all of us. The next morning the 103 was wrapped in tarpaulins. That night, while we slept peacefully behind bars in our dormitories, she was taken to the aerodrome ...

Nyukhtikov (an Air Force colonel) and his navigator Akopyan put on their parachutes. (The designers had by then been brought to the aerodrome, too). Concentrating totally on their mission, the airmen climbed into their places without a word. Probably, although they denied this later, they felt some apprehension as they prepared for their flight in this aircraft which had been designed and built in such weird circumstances – by a bunch of jailbirds, to put it bluntly. The engines were started and the 103 strained at the leash. Nyukhtikov raised his hand and the ground crew pulled the chocks from under the wheels. The aircraft was steered slowly towards the start line.

Andrei Nikolayevich walked steadily after her. Such was the man's natural authority that no-one ventured to stop or escort him. He cut across the field and we knew for sure that somewhere very near to the spot where he stopped and waited the aircraft's wheels would leave the ground. That was how it always was ..."

[49, No. 7, p. 43]

The 103 completed her trials without a hitch. It was decided that the type should go into production. That meant freedom! We waited for notification of our release – one day, two days, then a week went by. Then a directive came down from on high – the 103 was to be redesigned, and how!

One of the best points about the aircraft was that the crew members were not bunched together. There were two separate crew positions. Now it was decreed that to improve the reliability of the airframe the crew should be brought together in one cockpit. Only those who have worked in the aircraft industry know what such a redesign job entails. The task was all the more irksome in this case since the aircraft was successful in its original form.

The new model was dubbed the 103U. Fate was not kind to the modified plane. It flew well at its trials, although its top speed was 50 kph lower than that of the 103, but after a few flights disaster struck – the starboard engine caught fire. Nyukhtikov and Akopyan were again the crew. Nyukhtikov gave the order to bail out. Akopyan's parachute caught on something as he jumped and he was killed. Only burnt wreckage remained of the machine.

A new wave of arrests followed. It was not hard to predict what now awaited the jailbird designers. However, it was quickly discovered that the cause of the fire was a fuel leak from a pressure gauge tapping point. All the accused were released and the two fitters who installed the offending manometer were arrested.

The aircraft was accepted but did not go into production. It transpired that the high-altitude engines, around which the 103 was designed, were no longer to be made. A new directive came down: the aircraft was to be redesigned to accept engines of a different type, the production of which was now proceeding smoothly at a certain engine factory. The only problem with these engines was that they had proved unsuitable for use in any existing airframe and huge numbers of them were piling up at the works. Somewhere in the upper echelons a bright bureaucrat had spotted a way of killing two birds with one stone.

Tupolev was furious. Only a very radical redesign could adapt the 103 to take these unwanted engines and, even then, performance would clearly suffer. Moreover, all this extra work would take time. Feelings reached such a pitch among the zeks that Kutepov was forced to secure a promise from on high that as soon as the new plane flew the whole team would be released.

All this happened on the eve of the outbreak of war. "Vera" – the 103B, as the new model was called – was to be completed in Omsk, in an evacuated factory, and would make her first flight there. However, most of the members of the Tupolev DB were held in captivity until 1943.

"I was released on the 5th of May, six weeks before the war broke out," recalled Leonid Lvovich. "It was all thanks to my wife. I had, you see, never worked in CAHI before my arrest. I had no connection with CAHI. I worked in the Air Force's Scientific Research Institute (SRI) where I specialised in radio communication and radio navigation. When I was still quite young at the SRI I did a lot of work on the ancillary equipment for the ANT-25, the plane in which Chkalov and Gromov flew to America over the North Pole. Andrei Nikolayevich Tupolev, the plane's chief designer, was always dropping in to see it. He must have liked what I was doing because when they were getting a team together for the polar flight I was asked to join it and to take charge of the communications and radionavigation.

That was how I came to be on very friendly terms with Chkalov's crew – that is, with Chkalov himself and with Belyakov. With Baidukov, too, but I had known him before. Then there were Gromov, Yumashev and Danilin.

When I was arrested, my wife Yelizaveta Mikhailovna Shishmareva went out and fought for me. She wrote letters and submitted applications all over the place, declaring my innocence and requesting a review of my case. But it was all in vain. Then she turned to the airmen who had become Heroes of the Soviet Union. Remember that in those days Heroes were few and far between – unlike today's Heroes – and the whole nation knew their names. Two of these Heroes – Georgi Filippovich Baidukov and Sergei Alekseyevich Danilin – received her plea very sympathetically and wrote to the Supreme Soviet. My case was reviewed (if they bothered to review it, that is) and I was released.

Of course, I knew nothing about all this activity. Suddenly one fine day the governor came up to me and said: "Come with me." He took me to his office where I received the curt instruction: "Get your things ready." I could feel a cold wind blowing from the camps. Terrified, I ran to the Chief. The governor came after me but I managed to say to Tupolev: "Andrei Nikolayevich, they've told me to pack up and get ready to leave!" Tupolev slapped me on the shoulder and said: "Come on." That was all he said. Then I realised: he probably knew what was afoot.

I was taken to Butyrki prison. I was kept in a cell there for a while, then I was summoned to the governor's office. Kutepov was sitting there. He got up, offered me his hand and said: "Congratulations. The Soviet government has checked and established that there is no case against you and has decreed that you shall be released." Then he gave me a document, which I have kept safe to this day, which records that from such-and-such a date I was detained at Butyrki, spent a certain period in the North, plus a hell of a lot of other details ending with the date of my release – 5 May 1941. Meanwhile he was giving me his parting advice, explaining that it was not in my interest to fraternise with the prisoners, that we who had been released were not in the same category as those desperate criminals who remained in custody. Over-friendly relations with such types could lead to serious consequences. I concluded from this that I was being sent back to Radio Street as a free employee. They could hardly decide otherwise, as I was responsible for a significant part of the equipment on the "Vera".

Kutepov said his farewell but I did not move. That shows you how free I felt. I said to him: "But how am I to get home? I have no money. Could you let me have 15 kopecks for the tram?" 15 kopecks was the flat rate on the trams in those days. These men could not muster enough courage between the two of them to give me 15 kopecks out of their own pockets so I was once more led off to a cell and locked up. How I kicked myself! I do not know how long I sat in that cell; we were never allowed watches. Finally I was led outside and put in a pick-up truck. The gates opened one after the other and I was driven out to freedom.

At this point I became confused. I did not know which way to go. Yelizaveta Mikhailovna and I lived in a huge communal apartment on the Sretenka, opposite the "Uran" cinema in a rambling building which had once housed the "Rossiya" insurance company. Before my arrest we had two rooms there – a big one 24 sq.m. in area and a small one of 8 sq.m. When I was arrested the smaller room was sealed up pending further investigations. (There were so many sealed rooms in the communal flats of Moscow then!) Lisa had been left with just the one room. There were five other families in the flat besides us. What a fuss there would be if I turned up there! So I went to Petrovsky park to the house of Professor Shishmarev, Lisa's father. Lisa and I were not officially married although the children had been registered with my surname. Lisa's parents were at home. First the "oohs" and "aahs", then without further ado they rang Lisa. She grabbed our two little boys and in half an hour we were reunited. The younger did not know me from Adam. He was only five months old when I was arrested.

Two days later I was back at work, the only difference being that I was now a free engineer. I was worried about how my friends would receive me – after all, it had been declared that I was "clean" but they were not. It was an awkward situation. However, when I got there their attitude towards me had not changed. In fact, as soon as I turned up three of them – Sergei Mikhailovich Yeger, Sergei Pavlovich Korolyov and – I have forgotten who the third one was – came to me and asked me to track down their wives in Moscow and tell them everything. You see, our wives thought we were in Butyrki – at least those fortunate wives who were allowed to send parcels.

Years later, Yeger and I went to see Sergei Pavlovich at his dacha – a sturdily constructed house with two floors and a green fence. I do not remember what occasion we were celebrating. There is a museum there now. We sat there drinking cognac in the approved fashion. It was late one winter night. Sergei Pavlovich said suddenly: "You know, lads, there are times when I think that the door might just burst open and in will come the screws and they'll say: "Come on then, get your things ready!" Do you think that could possibly happen? Or not?"

We all three thought that it could. It was in the 1950s. Beria had gone by then but people were still frightened. We saw the world through the eyes of engineers. We saw a classic sinusoidal wave rising and falling. That is why we thought it could all happen again. No doubt it could not happen now.

We think we understand everything now. We assume that as far as Lavrenti Pavlovich Beria is concerned all the evidence is clear and unambiguous. Then, however, when he used to come to SDB-29 to see us – very obliging, very attentive – all that we saw was that he knew nothing about engineering. That came as no surprise, though. Maybe somebody had convinced him that we really were spies – how were we to know? At any rate many of our number dared to hope that Beria would try to see that justice was done. Some prisoners even attempted to "open his eyes to the truth". We were convinced that sooner or later the misunderstandings would all be sorted out."

Yuri Borisovich recounted how from time to time Beria used to arrange receptions in the prisoners' honour. The tables on the sixth floor were pushed together to form a horseshoe ready for the banquet. Standing in a lordly pose Beria would welcome his guests at the door of the dining room.

Yuri Borisovich recalled:

"An extraordinary table was spread for us, laden with caviar, sturgeon fillets and fruit. When all the guests were gathered Beria would stand up at the head of the table and start his speech. He would speak in an almost affectionate voice, as if to say: 'So, I want to ask your advice about where we go from here, how the work of the Bureau will proceed, whether we need new people and, if so, from what area of research. Let's forget about unpleasant things for today and have a good time. Today you are my guests, so relax and feel free'.

On one occasion, after Beria had confided thus in the prisoners, Bartini suddenly got up and walked towards him. I was taken aback. Bartini was my friend and associated with only two or three other prisoners apart from myself. He was very reserved. He walked up to Beria, lifted his handsome, proud head and said: 'Lavrentii Pavlovich, I have been wanting to tell you for a long time that I am not guilty of anything – that I have been imprisoned for nothing'. He spoke only for himself – it was not considered proper that anyone should speak on behalf of any group.

How Beria's expression changed! The face of the good-humoured proprietor became a face of triumphant rapacity. He stepped softly up to Bartini and
said: 'Signor Bartini, of course you are not guilty of anything. If you had been
guilty, you would have been shot long ago. But it was not for nothing that you
were put in prison. Otherwise you would be in your aeroplane and off to freedom. In your aeroplane and away!' And he gestured with his hand to show
how an aeroplane flies, even standing on tiptoe to make the plane fly higher.
We thought it was funny at the time and laughed the kind of belly-laugh you
hear in prisons. But at the table next day even Makhotkin sat in silence."

Yuri Borisovich sat at the same table as Roberto Bartini,
Karlusha Szilard and Vasya Makhotkin. Bartini and Makhotkin did
not get on well but it was to be their fate to suffer together in the
DSEAE, at Omsk and in Taganrog prison. Bartini, always reserved,
withdrawn into himself, had little to say and hardly ever entered into
conversation at the table. Makhotkin, however, a tease and a joker by
nature, was irrepressible:

"Karlusha – eh, Karlusha," Makhotkin would begin,

"Why did you come to Russia?"

"To give a haand."

"To give a haand – and have you ever read Dostoyevsky?"

"I haave."

"Well, if you haave, then how did you think you could help?
Roberto, have you got a son?"

"Yes," replied Bartini severely, not suspecting that he was about to
be tricked.

"Well then, tell your son that when somebody else's country is in a
mess, he should keep his nose out."

On one occasion Makhotkin produced a primary school textbook
called "Our Mother Tongue". (Where he had got hold of it always
remained a mystery.) In a loud voice he started to read from it a story
about a boy called Vasya who always dreamed of becoming a pilot. The
boy grew up, got into a pilots' training school – and his dream came true:
he became a bold and skilful polar aviator. There is even an island in the
Arctic Ocean named after this boy. The story ended with the words: "So

let's all be like Vasya Makhotkin!" "And Makhotkin's behind bars!" concluded the hero of the tale jovially.

Karl Szilard, the Hungarian mathematician, was among those pilgrims who hastened to Soviet Russia, wanting to live and work here. Szilard arrived with his wife and their tiny baby daughter. They lived in Leningrad and were happy. Now, in prison, he knew nothing about their fate. Any attempt to communicate with the outside world would have resulted in the severest possible punishment.

The most dramatic incident involving Szilard took place during the war when the Tupolev design team was evacuated to Omsk. In the factory there, where the prisoners lived in one of the workshops, there were no baths, so they were taken once a week to the city's public bathhouse. Usually it was late in the evening so that they were less likely to be seen. On one of these trips Szilard was left behind at the bathhouse. He himself did not remember how this happened. He only remembered the panic and the horror when he realized that his friends, the guards and the windowless van had all gone. He started walking up and down in front of the bathhouse in the hope that they would come back for him. But they did not come for him so he decided to make his own way back. He did not know the way. He could not ask because he had a marked accent which could be taken to be German. He decided to seek out a policeman (after all, policemen are supposed to protect people) and confess everything. But when he saw one at a tram stop he lost his nerve. What if he didn't understand? Then he was really in for it – they would think he was trying to escape!

So, he wandered around the town wondering what he should do. He decided to walk out to the edge of the city and then work his way around the outskirts. The factory was just outside the town – this much they knew. In the course of those wanderings he kept looking into windows – this gave him a real taste of domestic warmth. Suddenly, through a basement window, he saw his own wife. She was washing their daughter over a bowl. Then she wiped the little girl's face with a towel. He did not know how he found the strength to straighten up and carry on – but that is just what he did. Karlusha found the factory. Nothing happened to him. Not even the most trivial punishment. The prison authorities could, by this time, see fairly clearly what kind of person they were dealing with. But Karlusha spoke to nobody for two whole days. He hardly touched his food. Later, when he began to feel better, he told Rumer the whole story. They were very close.

Many years later, in the early 70s, when Szilard was back in Hungary and Rumer was in Akademgorodok, a certain Hungarian journalist asked Yuri Borisovich for an interview. He wrote:

> "Professor Rumer lives in Akademgorodok. He is renowned theoretical physicist, brother of the late O. Rumer, the translator of Petöfi's poems, about whom we recently published an article. Professor Rumer greeted me in brilliant Hungarian, quoting Kossuth and Arany. When I expressed admiration for his Hungarian he said that he had a Hungarian friend, and that they had drunk bitter wormwood together. I was puzzled. 'How do you mean?' I asked. 'Oh, just for a bet', he said, quickly changing the subject."

Yuri Borisovich spoke perfect Italian, too. Bartini can take the credit for that.

In 1922, at the time of the Genoa conference, an attempt by Savinkov's men to assassinate the Russian delegates was foiled by some Italian communists. One of the main organisers of the group which prevented the assassination was a young Italian Communist, "a strange aristocrat" whom the Fascists had long been shadowing, called Roberto Bartini. Bartini was indeed an aristocrat. He was the only son and heir of the wealthy Italian baron Ludovico di Bartini. When Mussolini's henchmen began to keep such a close watch on him that no disguise was of any use and even constant movement between Lake Como and rural Sicily did not shake them off, Roberto Bartini travelled to Petrograd on a false passport, presenting himself there as a "returning emigré", the son of an architect from Odessa. He was twenty-six years old when he arrived in that cold, grim winter of 1923 and settled into a small basement room in Moscow. In our encyclopedias Roberto Bartini is recorded as an outstanding Soviet aircraft designer, creator of a dozen experimental and research aircraft. He died at the age of 77, having lived in the Soviet Union for 51 years, for 45 of which he was a chief designer. Bartini's favourite saying was a line from the Koran: "Help us, O Allah, to avoid that which we cannot overcome and to overcome that which we cannot avoid."

The evacuation of all design teams and their factories got under way in the summer of 1941. The Tupolev team was sent to Omsk. Not long before the evacuation Korolyov arrived at the DSEAE. He did not have any long-winded stories to tell – just a phrase or two here and there.

The things he had been through! He had been "a-digging for gold" in the frozen earth of Kolyma. He had developed scurvy and lost his teeth. He still had headaches after a blow on the head. He had a wound which was hardly managing to heal. He had fought his "trusty" team leader in single combat ("A thug with the law behind him – now there's a man to reckon with!"). Then there was the time they were supposed to put him on a boat bound for Moscow. They missed it – and it sank without him! ("See how lucky I am?"). The general opinion of the inmates of the "golden cage" was that Korolyov had been through a worse hell than any of them. And all on account of that work on rocket technology which had been approved by Tukhachevsky and because of which Kleimenov and Langemak had been shot and Glushko had been imprisoned. Nevertheless, the first memorandum which Korolyov addressed to the prison authorities as soon as he got back from Kolyma was a report on the urgent need to expand rocket research.

Tupolev released Korolyov from the assembly line and allowed him to work in the design office on calculations for a rocket motor. At this time Korolyov enjoyed close relationships with Rumer, Szilard and other physicists and mathematicians. "Doing calculations" is a phrase used in aeronautical engineers' jargon but in fact it refers to genuine scientific work. Yuri Borisovich related how, when he was in exile and was able to get on with his scientific studies, he wrote several articles which were inspired by his discussions with Korolyov. As examples of these he mentioned, among others, his article (published in Reports to the Academy of Sciences 1949) on "Turbulence in Fluid Flow through an Annular Opening", then his "A Problem Involving a Submerged Jet" (1952) and "Convective Diffusion in a Submerged Jet" (1953).

In 1943 the decision was taken to proceed with rocket research. It is well known that a report to the Supreme Commander concerning the German "vengeance weapons" – the V1 flying bomb and the V2 rocket – played no small part in this decision. Korolyov was transferred to Kazan and the work on rocket motors was started in prison. A test rig was built and installed in a Pe-2. The flight engineer in charge of the test flights was Korolyov himself. Meanwhile, in the world outside, rocketry was still regarded as mere pyrotechnics. Even then (in 1945!) the national newspapers were printing articles in which famous aircraft designers warned of the harm done by rocket research.

1943 was a year of great achievements in the aircraft industry and in the "Great Release" Tupolev and almost all those on his staff list were set free. By that time the special design office at Omsk, which had grown substantially bigger than it had been in Moscow, had been divided into several independent design teams. In one of these, which was known as Experimental Design Office No 4, Roberto Bartini was the chief designer. Bartini's projects, being totally new, required the involvement of academic scientists. He was given Rumer, Szilard and Walter. Makhotkin also asked to be transferred to Bartini's team. They came back to Moscow practically all together, their arrivals separated only by short intervals. The Tupolev design office was re-established on Radio Street. In the building no traces of prison bars or prisoners' beds remained. The team was free.

Bartini's designers were assigned premises in the Rostokino area of Moscow. They were transferred to Taganrog in 1946.

In prison

Chapter 15

Teaching the Teachers

In 1946 Bartini's design bureau was moved to a new factory and was given a new assignment. There were 126 prisoners on its staff. They were only allowed to mix with the factory workers if accompanied by one of the "screws". Once a week they were driven to the town baths. Life was hard for everyone after the war and the prisoners shared the general hardship. They never had enough to eat. Nevertheless, it was here that Yuri Borisovich met his fate.

"How did it happen?" Olga Kuzminichna repeated the question.

"It was quite simple. I was just about to get married. My fiancé Ivan, worked at a nearby factory and as our wedding day approached he tried to talk me into getting a job there. He said the pay was better. I did not really want to move but I thought – all right, if that is what he wants I'll do it. I went to the shop manager to give my notice. He read my letter and frowned:

"So, aren't you Kuzma's daughter? Why do you want to change jobs?"

"Well, you see, I'm getting married and apparently the pay's better where my fiancé works."

"If it's a rise you're looking for you can stay here. We've got a new DB. I can fix you up with a job in it working for a professor."

He's joking, I thought, some chance of a professor coming here – they only exist in books, anyway. Nevertheless, I accepted. I was shown into a glass-walled office. I walked up to the desk. There was a man sitting there with his back to me. He turned round. I had never seen such sparkling black eyes. He started to get up and he seemed to go on and on getting up – he was so tall. Then he gave me his hand and smiled. "Yuri Borisovich," he said. "Olga!" I blurted out and grabbed his hand. I felt I could not let go of it. I felt my cheeks burning. And I still can't let go of this hand. I've been holding it for 40 years."

On the first anniversary of Yuri Borisovich's death the Institute Nuclear Physics in Akademgorodok arranged a seminar in his memory. Rumer's students and friends were all invited. Many of them came; those who could not come sent telegrams or letters. A very warm-hearted letter along with several pages of reminiscences arrived from Kazan – the writer was Mahmud Mubaraksheyevich Zaripov.

> "In the early spring of 1946 I was sent to work in the aircraft design bureau of Chief Designer Robert Ludvigovich Bartini. I was appointed to the team that was working on aircraft vibration. The head of this team was Professor Yuri Borisovich Rumer. His name was well-known in the mathematical physics faculty of Kazan University, where I had done my degree. Our lecturers there had urged us to read Yu. B. Rumer's monograph "An Introduction to Wave Mechanics" as well as "Notes on Atomic Physics" by Blackwood and Hutchinson, which had been translated by Feinberg and edited by Yu.B. Rumer.
>
> Knowing that Rumer was one of the USSR's most eminent physicists, I was a little nervous when I introduced myself. Yuri Borisovich first went through all the information on my forms and all my personal details. Then he went on to ask who had taught me and what I had studied. He did not seem to be completely happy with the answers I gave concerning my knowledge of physics. Then he explained what my duties would be as an engineer in his team.
>
> I became aware in due course that the team was not simply concerned with calculations relating to aircraft vibration. Chief Designer Bartini was a man with an enquiring and fertile mind. He looked far ahead into the future development of aviation. Yuri Borisovich was his right hand man and his scientific consultant. The work of the team, however, went beyond consultations and calculations – we broke new ground, too. Original research was done in the fields of vibration and aerodynamics as we sought solutions to the problems which arose in the design process.
>
> I fell into the swing of this work without too much difficulty and discovered, to my joy, that five years in the Taishetian taiga had not erased from my memory the knowledge of mechanics and mathematics that I had gained at the University of Kazan.

Not long after I started work with the team Yuri Borisovich asked me if I would like to do some theoretical physics. Naturally, I accepted his offer eagerly and immersed myself in "Field Theory", the book by Landau and Lifshitz.

I had plenty of time for these studies. We worked a 10-hour day. The DB was situated on the outskirts of the town on the Gulf of Azov. Next to the factory grounds was a wood which for some reason was called Quarantine Wood. During lunch breaks in the summertime we used to go down to the shore to swim and sunbathe. In our very first year there we were given allotments where we could grow vegetables. Sunday was our day off. The DB library was not very well-stocked. There were not many journals but at Yuri Borisovich's insistence we used to get a number of physics journals (including JETP and 'Phys. Rev.').

So I had what I needed in order to study theoretical physics and I got on with "Field Theory". Yuri Borisovich was a fantastic teacher. My study of Landau's course was accompanied by excursions into the history of physics and as part of my programme Yuri Borisovich included an account of his own research, too. It was not long before I discovered that he was working on a unified field theory. At that point in his life this was the centrepiece of his scientific thinking.

Yuri Borisovich suffered a lot because he was cut off from the scientific world in which he could have discussed the issues he was wrestling with. He came round to thinking that he could solve this problem by using me as devil's advocate. He would outline his approach to whatever problem he was working on and ask me to tackle him frankly and critically about any point I did not understand. He wanted me to object boldly if I spotted anything that I thought was incorrect. Then I would go through the topic again, explaining it as I understood it. These discussions took place in the evenings, after work. My role was that of hawk-eyed critic – I was to express any doubts I had about what he told me and ask as many questions as possible.

Yuri Borisovich strove to expound his ideas with the utmost clarity and the most rigorous reasoning. I served as his sparring partner when he was working on his unified field theory. In so far as I was capable, I played my part in the elaboration of a number of formulae and solutions. Yuri Borisovich worked indefatigably. He was an optimist. He had high hopes of returning to his rightful place in the scientific community. After all, the end of his prison term was approaching and he would not return to science empty-handed.

I remember how he once suggested that I should write a dissertation. He was most taken aback by my rejection of the idea and even more by the reason for it. I explained how hopeless the future looked to me. This had a terribly depressing effect on Yuri Borisovich – in fact I regretted having so thoughtlessly upset his mental poise. Nevertheless, despite my negative reaction he tried to make me see things in a more cheerful light. He was an amazingly well-read man and in defence of optimism he mustered a host of writers, even quoting the classic literature of the Orient. He was a polyglot; he could even speak several Asian languages. He knew whole passages from the Eastern poets by heart. He tried to learn Tatar (my mother tongue) and learned several Tatar songs.

Yuri Borisovich's closest friend in the DB was the head of the aerodynamics section, the mathematician Karl Szilard. He spent a lot of time with Bartini, too. He loved to encourage young people with talent.

The time went by ... The end of Yuri Borisovich's sentence approached. Unfortunately he could not go back to his beloved Moscow. Instead he was exiled to a settlement in Krasnoyarsk region. I was horrified when I thought about his future (I knew what life in the taiga was like).

When I finished my own sentence I, too, was sent to a settlement in Krasnoyarsk – it was in the Buguchan region. I got hold of Yuri Borisovich's address. I wrote letters to him and he wrote back. On my way back to Kazan from exile in 1955 I stopped at Novosibirsk and called on Yuri Borisovich and Olga Kuzminichna. They were making a new life for themselves, which I was very glad to see. Yuri Borisovich wanted me to study for a higher degree under his supervision but I had my mother and grandmother waiting for me in Kazan, completely dependent on me. Apart from that I was already married and our first child was on the way. I had to get to grips with my life."

In 1964 Yuri Borisovich was to be the "opponent" on the examining panel when Mahmud Zaripov defended his Candidate's dissertation. In 1981 a delighted Rumer was to write an assessment of Zaripov's doctoral thesis. Zaripov's letter ends with the words:

"In everything I have tried to follow his example. All the joy and happiness in my life is due to that fact. I am eternally grateful to him!"

In the summer of 1947 Rumer was summoned and informed that an order for his early release had been received. He was sent to the city prison escorted, as was normal in such cases, by two guards. That was normal procedure – on the day of his release a prisoner would not work; instead he would be escorted to the prison where his release papers would be prepared. The next day he would return to work as a free employee, receiving a salary now instead of his prison rations.

However, Yuri Borisovich came back to the special prison in the evening of that same day, still escorted by his two guards. In the telegram ordering the release the name of the prisoner had been misspelt – two letters had been transposed. Someone else was being released, not Rumer. Of course, his fellow-inmates did not know this and next day people kept coming to congratulate him. Every time someone came he got up from his desk, straightened up to his full height and said: "No. There has been a mistake."

The last year of his sentence was probably the hardest for him to bear. Yuri Borisovich did not recall any particular details, nor did he talk explicitly about the vulgar behaviour and filthy language of the "bosses" – he did not like to repeat such things. However, he could both make and take a joke. On the occasion of his sixtieth birthday (which he reached during his time as Director of the Radio-Electronics Institute in Novosibirsk) his students gathered together and first read out, then presented him with an "Extract from the Decree concerning the Appointment of Yu.B. Rumer as Director of the REI" It reads as follows:

> "We are sending to you an alumnus who has served with us over a number of years.
>
> *A portrait in words:*
>
> (1) Looks – intelligent
> (2) Dressed – in the appropriate uniform
> (3) Face – deserving a medal
> (4) Profile – could be minted
> (5) Outlook – prejudiced
> (6) Temperature – beyond permissible norm
> (7) Distinguishing marks – five-dimensional

Personal details:

(1) Origin – legitimate
(2) Education – secondary education at Göttingen (Three years at school in Göttingen and 10 years of practical work in Siberia.)
(3) Religious affiliation – Symmitic
(4) Knowledge of languages Turkish (reads with help of dictionary) Ugrian (can make himself understood) GULAG (fluent)
(5) Scientific publications – in the care of the State
(6) Government awards and incentives – clean sheets and a packet of cigarettes.

Supplementary information:

(1) Character – Sofetsticated
(2) Favourite shapes
    (a) rings (benzole)
    (b) grids (Isingian)
    (c) female
(3) Favourite numbers – 58.20
(4) Attached to – matrices, terriers, Pfaffians
(5) Hobby – transmission of genetic information
(6) Recommendations as to regime – only transitions of the second kind to be permitted

Conclusion:

Will serve well as Director of Institute.

The final section may require clarification. Each point, except the third, reflects part of Yuri Borisovich's wide-ranging scientific activity. The first refers to his fruitful collaboration with A.I. Fet (apart from their publications relating to original research they also jointly wrote two monographs: The Theory of Unitary Symmetry" (1970) and "The Theory of Groups and Quantum Field" (1977).) The fifth point refers to Yuri Borisovich's amazing work "Concerning the systematisation of Codons in the Genetic Code", which consisted of only two pages. Yuri Borisovich wrote this paper when he was 55 years old and was as delighted with it as a child with a new toy. The paper stimulated a record number of responses – several hundred, in fact. Postcards even came from the most exotic parts of Africa. Not long ago a young biologist who never knew Rumer said that if a scientist wrote nothing else in his whole life except this one little paper he would be remembered by all future generations of biologists.

As regards the third point, the reader will guess correctly that this refers to the article of the Criminal Code under which Yuri Borisovich was found guilty in 1938. It also refers to the fact that, ironically, his home telephone number was 58–20. In fact his family still use the same number today. "The security organs are especially fond of me," Yuri Borisovich used to say, "they even installed a hot-line telephone with a privileged number."

On 26 April 1948, exactly ten years to the day after his arrest, Yuri Borisovich was ordered not to report for work. There could not be any mistake this time. Once again he was sent off to the city jail with two guards. However, he did not turn up at work the next day. He did not turn up on the day after that, either. The prisoners began to get worried.

"Before Yuri Borisovich was released I found a room for him," related Olga Kuzminichna, "just as I had found a room for Karlusha a month before and for Robert Bartini a year before that. Although we had decided to get married and my parents had got a room ready for us complete with nice little curtains and pillows, although the food for our wedding reception had already been bought, nevertheless he had to live in a different place for a while. It was Holy Week, when no marriages take place. We planned to get married the week after – but he did not come back from the jail.

I went there to find out what was happening. When I asked about the prisoner who had been brought from the special prison the officer on duty told me that a group of prisoners was being transferred and that they were setting off at any moment. No information could be given that day. I should come back the next day. As I stood outside by the wall I could hear the bark of dogs. A tram came along at that moment and I quickly jumped on to it. If I had stayed ten more minutes at the gates I would have seen him. Perhaps it was better, though, that I did not stay. You know what I mean – when the dogs are barking and there are sentenced for 25 year in handcuffs walking at the front of the column and then all the ordinary transit prisoners following behind, it is not the most pleasant sight. I had seen just such a scene before.

We only found out what had happened when he got to Yeniseisk. It turned out that on the 10th of March 1948 a secret decree had been issued to the effect that persons convicted under Article 58 sections 'such-and-such on completion of their 10-year sentence should be sent into exile for such-and-such a period. That meant that if you were an Article 58 offender, whether or not your sentence included loss of rights, you were given your marching orders. The last of our people who got out before this decree came into force was Karlusha. He was released on the 8th of March. The next to be released was Yuri.

When he still did not appear at work on the 27th of April I decided to go to Moscow immediately. I had to take "Pentoptics" with me. Yuri had given it to me earlier and I had smuggled it out of the factory and hidden it at his request with Karlusha Szilard. By the 2nd of May I was already in Moscow. I was staying with the Rumers when a telegram arrived at the beginning of June:

"arrived Yeniseisk send thousand".

I took Osip Borisovich in my arms and started to dance a waltz around the big square entrance hall.

"Yeniseisk (no date)
My dear Olenechka!
I have been missing you dreadfully and was afraid for a while that I had lost you. Life seemed completely pointless. Then your telegram arrived and I was overjoyed, especially as it arrived at the very moment when I learned that I had a

chance to build some kind of life. I have got myself a job. I am to be head of a department in the Teacher Training Institute ... Apart from all that, I hope that you have got my papers with you. If Dau is not in Moscow (why have I not received his telegrams?) then it is essential that my work should be given to Leontovich. Ask him to write an account of the results for a preliminary report. You know how much this work means to me. I hope everything is going well for Karlusha and that my papers are safe. If not, tell me by telegram and I will get on with rewriting them immediately.

I was thrilled to receive so much money. You see, on the journey I met some people who shared with me the last that they had and I have been able to pay my debts straight away ..."

This was Yuri Borisovich's first letter to Olga Kuzminichna from Yeniseisk. The journey to Yeniseisk had taken more than a month.

"I did not know where they were taking me." said Yuri Borisovich. That is one of the unpleasant things about being a prisoner in transit. The transport of prisoners is, in fact, a peculiar business altogether. You are packed like sardines in the trucks, there are all kinds of people all mixed up together – in a word, everything is done properly. However, a man like myself, who has already done his ten years, feels like a king in transit: the guards leave him alone, the old lags do not harm him – they reckon that a man who has been in prison for ten years has proved himself.

As this trainload of prisoners made its way from stop to stop (with a monstrous lack of urgency!) more and more newcomers were shoved into our already crowded trucks. Somewhere in the Urals a little old man joined us. He was a professor, a complete invalid, paralysed all down one side. He asked me for a smoke. I told him I had no tobacco of any kind, that I got my smokes from the lags (the lags were always bringing me bread, tobacco etc) and that I would ask Ashot to help him.

Ashot was an Armenian trusty who was travelling in our truck. He had great respect for me and was always trying to convince me of the superiority of the social institutions accepted among thieves compared with those of conventional society. ("For example, you vote and make your decision according to the will of the majority – that's wrong. The way we do it, even if one man is against it the decision will not be taken. What if just this one man has seen things in the right light?")

I asked Ashot to feed the old man. "Let's have a look what kind of bloke he is," said Ashot. When he had made the old man's acquaintance and they had chatted for a while, Ashot said: "He's OK." The old man ate greedily and then started to apologise: "Sorry, Ashot, I'll be the ruin of you at this rate." "No," replied Ashot, "Nobody can ruin me. The State has had five goes at it and got nowhere." Ashot and I parted at Krasnoyarsk. I was taken out of the truck there and after that was on a boat for several days. Ashot stayed on the train and travelled on."

Yuri Aleksandrovich Starikin takes up the story:

"I worked with Yuri Borisovich in Yeniseisk from 1948 to 1950. The rector of the institute called me into his office and told me that a certain Professor Rumer was being sent to work with us – did I know him? I had never had occasion to read any of his major publications; I knew him only as the co-author of a dictionary of physics. I was then head of the department of physics and mathematics at the Teacher Training Institute and also secretary of the Party branch organisation.

I first met Rumer in the rector's office. In came a man dressed in padded jacket and trousers, worn out after his long journey. He thought it was simply luck that had brought him to this interview but, of course, it was all planned. When the state authorities sent academics like Rumer into exile they "took care" regarding their employment. So even before Rumer arrived our local KGB chief, a well-educated and astute man named Grin, had informed our rector that Professor Rumer was on his was to us and was to be employed as a teacher of physics.

At that time institutions could not simply give jobs to ex-prisoners as they saw fit. Later, when Yuri Borisovich first arrived in Novosibirsk from Yeniseisk he called in at the teacher training institute to see the director, Sinitsyn. The latter, on learning that his visitor was a professor of the University of Moscow, spoke to Rumer very respectfully and was delighted to offer him a post as head of department. Rumer accepted and the director asked to see his documents. Instead of a normal internal passport Rumer produced a paper stating his right to reside in certain specified areas. Sinitsyn rose, pointed to the door and said: "Leave the premises!" It was surprising that Yuri Borisovich was never resentful about anything or towards anyone.

By 1954 he was head of the department of theoretical physics in the Siberian section of the Academy of Sciences in Novosibirsk. I went to see him at his office. I was in difficult financial straits at the time and needed to find some extra work. Yuri Borisovich went to this very same Sinitsyn and talked to him as if nothing had ever happened between them. Sinitsyn gave me a job.

The only other exile at the Yeniseisk Teacher Training Institute apart from Yuri Borisovich was a language teacher, an ethnic German from the Volga. All through the war he had been an interpreter at the front and had worked at a very high level in the military hierarchy. When he was demobbed he was told that in recognition of his services he was free to settle anywhere he chose. He declared that he only wanted to be with his wife. His wife, also an ethnic German, had been exiled to Yeniseisk. He arrived in Yeniseisk where the authorities immediately registered him as a resident, took away his passport and issued him with the "Ausweis" carried by exiles.

Most of the exiles in Yeniseisk were, in fact, Volga Germans and most of our students were the children of these German families. When they finished their courses they were transported like prisoners to the places where they were directed to work.

The other section of the town's population – the original inhabitants – tried not to have any dealings with the exiles. Not that the exiles were all that keen to establish closer relations, which were in any case frowned on by the authorities. Although Grin did not do much to keep the exiles separate they kept themselves to themselves, thinking that thus they would avoid making matters worse. After all, every one of the exiles was convinced that there had been some kind of mistake, that soon everything would be sorted out and they would be able to go back to their former lives.

However, there was an occasion when the exiles broke their own unwritten rule. When Beria was toppled in 1953 his daughter was exiled to Yeniseisk. Her husband was a permanent way engineer, the manager of the Transcaucasian railway. He lost his post because of her and applied for a divorce. The divorce case which ensued created a lot of fuss, partly because she was blind. The case horrified the Yeniseisk exiles. They were all most sympathetic towards Beria's daughter and tried to help her in any way they could.

It is worth reporting another thing that happened at Yeniseisk. The Deputy Minister of Education once came on a visit. The first thing he did was to hang up a sign: 'The Deputy Minister is available daily between 7 and 9 to examine complaints and requests of any description.' He was in

Yeniseisk for a week and a half. In all that time he did not hear a single complaint. Nobody went to him to complain. The day he left he told us that the Ministry had been considering closing the Institute but now he had realised that this was out of the question; quite the opposite, they should be talking about giving it a higher status: 'You are turning out the cadres we need, people who will work in remote areas of Siberia, reliable people who will not drift off elsewhere.'

All the lecturers were newcomers. I myself had come from Leningrad. In my youthful enthusiasm I had decided to go out and see the world. I had just got married and my wife and I decided to see the world together. That is how we came to be in Yeniseisk.

There were mainly single-storey wooden buildings, a few built of brick, a sports ground with a bare earth surface. There was no water supply; water was brought in in big tanks. The wemen used to pour it into their own smaller drums.

Spring used to come to Yeniseisk just before the First of May holiday. The river flows north and when the ice began to break up huge piles of it accumulated on every bend in the river. Not far from the town the Yenisei sweeps round in a huge loop and there a monstrous dam of ice always built up. The water rose so quickly above this dam that it would overflow the river banks and flood the houses in Yeniseisk. It would come right up to the institute. The institute was housed in a two-storey stone building but even so classes were cancelled until the floods abated.

The winters cannot be described as hard, although the temperature did get down to minus 50 and below. I lived there for nine years and those were good years for me. There was an excellent group spirit and I got on well with the people. It was there that I got to know Yuri Borisovich. I worked with him later in Novosibirsk.

Yuri Borisovich took his work in the Teacher Training Institute very seriously. He was always looking for the gifted ones among the students. There were some good kids there; Misha Kirillov was one. Yuri Borisovich gave him one-to-one coaching and hoped to make a great physicist of him. Unfortunately, Misha became ill and died.

Yuri Borisovich did not want to get me into trouble so he did not initiate any contact between us on the personal plane. Where work was concerned, however, he gave me a lot of help. Later, in Novosibirsk, he was able to relax and make friends and there we became very close. Once when he was already director of the REI he asked me: "Yuri Aleksandrovich, what do you think, would they let me join the Party?" The question as to whether to join or not did not arise in his case."

On the 22nd of June 1948 Yuri Borisovich wrote to Olga Kuzminichna:

"My dear Olenechka,

Since the day I heard that you are going to be with me I have been so uplifted and so certain of eventual success that my hopes have given me wings. We already have a two-room flat only three minutes' walk from my work. There is electricity, firewood for the winter, beds and a table. The institute has provided all the furniture. I am enjoying the luxury of a bed with a blanket, a pillow and sheets. If you want it the institute will give us a loan to buy a cow, which would cost 3000 roubles here. We have a supply of potatoes for the winter, too. I have been given a very warm and comradely reception and have been accepted as one of the group.

Therefore, as soon as you hear that Kaftanov has confirmed my appointment, leave Moscow and come to me... On the 5th of July our assistant professor of physics, Yuri Aleksandrovich Starikin, will arrive in Moscow by air. He will tell you absolutely all you need to know about the town and about what you need here.

They will supply me with felt boots for the winter and I do not need anything else. Of course, I would be happy if you did not have to work. I would like you just to look after the household and go to the English language class. We shall make raspberry and bilberry jam...

I am very concerned as to whether you managed to get my papers to Moscow. You know how important they are to me personally and not only to me. If Karlusha has got them and all is well with him then he should send them off at once.

Find out the addresses of two physicists who, I think, were very fond of me: Pomeranchuk and Markov.

My papers worry me more than anything else at the moment. I hope everything else will work out all right.

I like the countryside here much more than the South. I hope very much that I shall soon feel stronger and get over all that I have been through ..."

As the reader can see, Yuri Borisovich wrote this letter on the 22nd of June. On the 15th of June the "Journal of Experimental and Theoretical Physics" (JETP) had acknowledged receipt of an article "Towards a Theory of Magnetism in Electron Gas" by Yu. B. Rumer. There is nothing about pentoptics in this article. "Pentoptics" had still to be found or rewritten. In fact, Yuri Borisovich was to rewrite it. He was able fairly quickly to put together the basic outline once again but before submitting it for publication he naturally wanted to discuss it with Landau.

> "12/7/48
>
> Dear Rumchick
>
> The only reason why it has taken me this long to write to you is because I am, as you know, no good at writing. First of all my reply to your letter. We have already sent off your first article for publication ... As for your major series of articles, I think it would be best if you could come and talk about them in person; then you could decide what to publish and where. The idea of regarding action as the fifth coordinate is, of course, not one that appeals to me, my peculiar character being what it is. However, as you know, my views are far from being generally accepted.
>
> I have arranged for some money to be collected. Obviously Zhenka has written to you about the business with the Ministry of Higher Education.
>
> I will not write anything about myself because there is too much to write about. We can talk about all that has happened when we see each other.
>
> That is all for now, I clasp your hand firmly.
> Dau"

(The "Zhenka" mentioned by Landau in this letter is Yevgeni Mikhailovich Lifshitz)

In January 1949 the JETP published the first of Rumer's articles on pentoptics – "Action as a Coordinate of Space. Part 1."

On the 14th of February Yuri Borisovich wrote to Tanya Martynova:

> "My dear Tanyechka,
>
> I was very touched and glad to receive your letter; it arrived at one of the most important moments in my life ... "Action as Coordinate of Space. Part 1" was published in the January issue of JETP ... I feel I have achieved my life's main purpose and can now consider myself very happy, despite the various inconveniences of my present situation.
>
> How is your personal life getting along? How did you spend the war years? Were you at the front? What do you see now as your main aim in life?

My experience has convinced me that external circumstances, however difficult they may be, have relatively little effect on a person's state of mind. I have seen a lot of people who have gone through life without a hitch and yet have been very unhappy.

During the years since we last met I have known several people with whom I have formed close attachments and I have never felt lonely. I have been able to work flat out, ten hours every day year in, year out.

A big kiss from me as I squeeze your hand,

Yura."

After the publication of Part 1 of "Action as a Coordinate of Space" Parts 2–10 of the series followed. It took about four years for all ten articles to appear in print, not because they were still being written – the work had all been completed during the previous decade – but simply because JETP could not publish them any faster.

During these same four years Yuri Borisovich also published about ten other articles on a very wide range of topics in theoretical physics. For example: "Towards a Thermodynamics of Bose Gas"; "Towards a Theory of Electrical Conductivity in Metals in a Magnetic Field"; "A Study of a Submerged Jet" etc. It seemed that there was no area of physics in which Yuri Borisovich had no contribution to make at that time.

Yuri Borisovich's friends did all that they possibly could to help him and tried to obtain permission for him to leave Yeniseisk. They wrote letters, not tearful or pitiful missives but ordinary letters such as friends write. For example, Igor Yevgenyevich Tamm wrote:

"Moscow, 7/5/50

Dear Yuri Borisovich!

... I know all about your troubles, having met Dau and Mikhail Aleksandrovich (Leontovich) several times. I have spoken to Vavilov about you and I dare to hope that his intervention may help you.

It is amazing how actively and creatively you are working in so many different areas of physics. However, you are right to feel that I am sceptical about those aspects of pentoptics which you hold most dear.

I concede completely that your treatment of action as the fifth coordinate makes possible a most elegant formulation of the problem of the movement of a point in any given gravitational and electromagnetic fields. However, I usually find that I cannot follow you when you move on from new formulations of familiar equations to new generalisations. These do not strike me as "zwangsläufig" (self-evident, logically compelling) and, besides, I am sure (although I am keenly aware that I may be mistaken and will be glad to abandon my views if they are proved to be erroneous) that the development of theory in physics will follow different paths. Permit me to attach a few remarks about the first paragraphs of your work on pentoptics ... (Here follow three pages of detailed discussion of several questions which bothered Tamm.)

... Dear Yuri Borisovich, I hope that you do not mind me writing to you frankly about my doubts and misgivings; I do not want you to think that my rejection of your five-dimensional approach to a unified theory is prejudiced or without foundation.

I think that the most interesting of the new developments in science has been the discovery, as theoretically predicted, of the neutral meson with a life-span of $\tau < 10^{-11}$ seconds (it decays into two photons). Several colleagues think that the experiments carried out so far are insufficient to allow any talk of a definite discovery but personally I am not in doubt.

The report is published in the "Phys. Rev.", which probably does not reach you. However a synopsis of it will soon be appearing in "Progress of the Physical Sciences".

I was very pleased to hear about your marriage and the birth of your child. I am at a different stage altogether – I have a grandson and a granddaughter.

With sincere best wishes
Yours, I. Tamm"

Most of the letters that came were from Landau. Some were long, some short. They may not always have been coherent but they were always sincere.

"25/6/50

Dear Rumchick

You will get all the facts from Olya. She is great and has made a terrific impression. Your note about the rotation of the plane of polarisation is wrong – for the energy you write ED^* instead of $ED^* + E^* D$ (by the way, to be strictly correct it should be $\int (EdD^* + E^* dD)$)

I clasp your hand firmly. Dau

P.S.: For God's sake let Olya handle everything. Sorry to have to say this but where muddle is concerned you are no better than I am. So let Olya write the business letters. I cannot make head or tail of yours."

"8/10/50

Dear Rumchick,

It is very sad that everything has turned out for you as it has. I must say that everybody here has put in a terrific effort (especially Leontovich, just lately) and even Sergei Ivanovich has done more than could be expected – but the problem has turned out to be a lot more intractable than we could have imagined. ... Your Olya is great. She is a real 'Russian woman' who does not lose heart even in the most difficult situation.

Firmly, I clasp your hand. Dau"

The situation was indeed difficult. All the attempts to get Rumer out of Yeniseisk, to shorten his five-year term of exile, seemed to have failed. Nevertheless, by the end of 1950, thanks to the intercession of Sergei Ivanovich Vavilov, Rumer was transferred to Novosibirsk, where the Western-Siberia Office of the Academy of Sciences was then located. The plan was that the Siberian Office should serve as an intermediate step on Rumer's way back to Moscow. However, Sergei Ivanovich died in January 1951. Yuri Borisovich had not had time to complete all the formalities and was left without a job.

"For 30 months I was unemployed and for 30 months I was supported by my friends," said Yuri Borisovich. Rumer could not secure a post at the Siberian Office or anywhere else. He tried to get a job at one factory, then a second, then a third. Everywhere he went they were glad to have him. They would show him his desk and say that probably it would not be convenient for him to come to work at eight in the morning. If he liked he could come in at ten. However, all these offers ended in the same way – he did not in the end get the job.

"We started to lose hope of ever finding work," recounted Olga Kuzminichna. "We were renting a room in the flat of a certain lady called Galochka. It was all we could afford. It was a very small basement room. We had a mattress that stretched from the window to one wall and opposite was the wall of the stove which Galochka should have stoked from her side but never in fact lit – she managed with some other stove in her own room. Our wall used to be covered with ice in the winter. We never used the kitchen. We hardly did any real cooking – we just went hungry. Our suitcases served as our table and there was a plank fixed to the wall that we kept books on. Then by sheer good luck someone found us a lovely flat – two rooms, a kitchen, a stove for heating. We had it all to ourselves (the landlady had got married and gone to Riga for an indefinite period) and, most important of all, we only paid the official rent.

We kept on meeting people who wanted to help us. Klavdia Petrovna Chernova was one. She was the chief librarian at the provincial library. It was thanks to her that Yuri Borisovich was able to earn his first money in Novosibirsk. She paid him for some translations he did and then entered someone else's name in the records. The risks she took! But what the risks took Tanya Martynova! She did her best to be everywhere and help everybody. Fate had treated her cruelly. When the Germans reached the outskirts of Moscow she sent her son to the country for safety and stayed to help in the defence of the city. In the village to which he had been evacuated the little boy fell ill and died. He was about six years old. Tanya spent her whole life looking after people. She befriended Asmus and Pasternak and also took Anna Akhmatova under her wing.

At the time when things were at their worst for us she had equipped an expedition to Kuznetsky Altai – she was a geologist, you see. She took one truck from the geological party, brought it back to Novosibirsk and hired it out to a collective farm during the season when most of the farm work was done. The driver did all the seasonal jobs required of him and was paid partly in money, partly in kind. Out of that transaction we got cabbage, carrots, potatoes, millet, money and firewood.

Meanwhile Yuri Borisovich was still having to go to the NKVD every fortnight to report to them and extend his Ausweis. Every time he went it was as if forever. He used to take 100 roubles with him not knowing whether he would come home or be sent to a lumber camp. So we said farewell to each other twice every month. Sometimes he would be back in 15 minutes, sometimes the officer who happened to be on duty would just keep him waiting. He would not say a word – just close the hatch as Yuri Borisovich walked up to it. Then Yuri Borisovich would be waiting there until the evening, unable to complain. But even towards such petty tyrants he never expressed any resentment.

How he used to love to use the word "Comrade"! Near the Commune Bridge there used to be a bathing beach and we used to go there in the summer time. One day a military man got changed next to us and asked us to keep an eye on his things. Yu. B. stood up straight, he always had a very straight back anyway, and said delightedly: "Have no fear, Comrade ... (Yu.B. then named the man's rank, which he had worked out from his shoulder pips – he knew all about such things) everything will be all right, I'll look after your things!"

Yu.B. emphasised the word "Comrade". It was not "comrade" as in "Comrade, where do you think you are going? What are you up to?" but really "com-rade". "Com-rade" says it all – it means someone next to you, a shoulder to lean on, a helping hand. And we held out and survived thanks to our comrades. Yu.B. and I only remembered those hard times when we remembered our comrades."

It is true – neither Yuri Borisovich nor Olga Kuzminichna liked to remember that time – those 30 long months.

"What is the use of remembering?" Olga Kuzminichna said.

"Whatever I tell you it will all be misleading. There are only two things that are certain: one, that it was very hard and, secondly, that we held out thanks to the kindness that we met all around us. There were always people there who were ready to help. That is for example why we gave Misha his name – in honour of Misha Leontovich."

Rumer with his five-dimensional space

Chapter 16

On Pearl Street

In December 1952 the Academy of Sciences organised a discussion of Yu.B. Rumer's contribution to science. Yuri Borisovich had not been in Moscow, the city where he had been born and brought up, for almost 15 years. Now he arrived as an ex-convict stripped of his civil rights, carrying a temporary pass rather than a proper passport.

The result of the discussion was that "It is recommended that Rumer Yu.B. continue with his scientific research." In July 1953 he was appointed head of the department of engineering physics in the Western Siberia Office of the Academy of Sciences. In the September of that year the Ministry of Culture restored Rumer's rights and titles as Professor and Doctor of the physical and mathematical sciences with an unbroken record stretching back to 1935. Full rehabilitation followed in 1954.

Now in charge of a department, Rumer literally swamped his staff with work. Once again, naturally, he began to give lectures and seminars. Once again the young people flocked to listen to him.

He and Born resumed their correspondence.

"29 January 1955

Bad Pirmont West Germany
Markardstrasse, 4

Dear Rumer

Schönberg has sent me an English translation of your letter addressed to him of 31 December 1954 and also your article "The Optical-Mechanical Analogy". I was very pleased to have news of you after such a long break and I am glad to hear that you occupy a senior post in the Eastern Office of the Academy of Sciences.

It was with great interest that I read your account of your work, especially your five-dimensional approach to relativistic mechanics on which you worked 20 years ago when you were with me. I am afraid, however, that I am now too old to study such interesting ideas in detail.

About two years ago I reached compulsory retirement age – i.e. 70 – and had to leave my post at Edinburgh, where I had spent 17 years. Although we had grown to love Scotland and the Scottish people we preferred to spend our last years in our own native land. We picked a quiet little spot near Göttingen. I have sold the greater part of my scientific library and so I am hardly able to do any kind of work, except for my own amusement. Last December I was awarded a Nobel prize for my work on quantum mechanics which I published 28 years ago. This gave me enormous satisfaction. I am still thinking about the problem of determinism and chance in physics.

I hope to hear from you again. With best wishes from myself and from my wife, who remembers you well.

Yours, Max Born."

In 1957 the department of which Yuri Borisovich was head was expanded to form a new Institute of Radiophysics and Electronics. Yuri Borisovich became the Institute's first director. At first the Institute occupied two floors in the laboratory block of the Western Siberia Section. Then a very adequate four-storey building was erected for the Institute on Michurin Street. The ground floor housed a polyclinic full of lady doctors and thus it was that a good number of the bachelors on the Institute's staff acquired wives.

In Yuri Borisovich's list of publications there is only one entry for 1958 and that is a popular work – "Relativity in Time"– which he wrote in collaboration with Landau. This was the time when Yuri Borisovich was totally immersed in the organisation of the Institute and in looking after the young people whom he was gathering around him. There was no time for his own research. Even in a very short letter of greetings sent to Max Born to mark the latter's 75th birthday Yuri Borisovich's emphasis is on his young people:

"Dear Professor Born!

I am now several years older than you were when I was fortunate enough to be your student. I now have my own young people around me and I try hard every day to be as kind and helpful to them as you were to us. I learned this from you, my dear Professor Born.

Your devoted friend,

Yu. Rumer."

In 1957 the Council of Ministers of the USSR approved the decision to create the Siberian Section of the Academy of Sciences of the USSR. 30 km from Novosibirsk work began on the building of Akademgorodok. On 1 January 1959 the Institute of Radiophysics and Electronics became part of the new Siberian Section. In 1961 the Institute, or more precisely its theoretical wing, moved out to Akademgorodok. In the same year Yuri Borisovich reached the age of 60. He received an affectionate letter from Born:

"Markardstrasse 4
Bad Pirmont

Dear Rumer!

News of the approach of your 60th birthday has reached me. My wife and I wished to seize this opportunity to send you our heartfelt good wishes. May your work be crowned with success and your life filled with joy and happiness. We always look back with great pleasure on the time when you were with us in Göttingen. I remember, too, how you and I worked on the theory of elementary particles and although it was too early for all that and we did not achieve success nevertheless the work itself was interesting and great fun.

Tell us sometime a bit about your life. We do not even know whether you are married and we do not have the faintest idea about how you live. We have got a

nice little house in a health resort. You must come on a visit and see for yourself
sometime. We ourselves are too old to travel around much.

Sincere greetings and good wishes from my wife.
Your oldest friend.

Max Born."

The remarkable thing about this letter is that although they had
been writing to each other for six years by this point this was the first
time that Born asked Rumer about his personal life. Born was surprised by Rumer's answer to this letter and by his later letters. In his
commentary on the correspondence with Einstein, Max Born writes
about his assumption that such an ordeal as Rumer had been through
would have turned him against any system. However the letters which
he now began to receive from Rumer did not support this assumption.
Born notes in amazement that: "His sufferings did not embitter him,
did not make him hostile towards the regime or towards any person.
Quite the opposite, he wrote long letters in which he tried to convince
me of the superiority of the Soviet system not only on the political level
but also as regards morality" [1, p. 102]

Yuri Borisovich was given a second-floor flat at no. 10,
Zhemchuzhnaya Ulitsa (Pearl Street). This flat served as his living quarters. Two more flats – one with two rooms and one with three – were also
assigned to him – to house the Institute. These flats were on the ground
floor of the same block, at the bottom of his staircase. The three rooms
plus kitchen of the larger of the two ground-floor flats provided working
accommodation for the theoreticians – one room being equipped as a
laboratory. Yuri Borisovich's office was the smaller room in the two-
room flat, the larger one being the seminar room.

The Institute worked on a series of important scientific and technical tasks. For example, in the field of electronics the Institute not
only did theoretical work on ultra-high frequencies but also designed
and built new electron-wave instruments. Yuri Borisovich's colleague
N.I. Kabanov received a then first Soviet licence for his discovery of

the so-called "Kabanov effect". The Institute built the USSR's first gas laser. The "Chronical of the Siberian Section" records: "The Institute ran a permanent seminar on quantum optics which was of prime importance in the development of laser research and in the creation of a Siberian school of theoretical physicists specialising in quantum electronics."

In the theoretical wing and the theoretical seminars Yuri Borisovich found the work that was closest to his heart. It was a friendly, young and cheerful team. Director and subordinates were on comradely terms. Anybody could interrupt the director, show him the errors in this thinking (well, at least they could try) or make a joke at his expense. It was here that the custom arose of shortening Yuri Borisovich's name to its initials – he was long to be known simply as "Yu. B".

On one occasion the theoretical staff made so bold as to slip an extra sheet into the folder where the director kept lists of directives:

"...07. S.K. Savvinykh to be reprimanded for wearing a university badge. Reason: erudition should be written all over a person's face, not on his lapel.

   08. F.R. Ulenich to be required to take off his army boots before jumping over chairs. Reason: because I say so.

   09. My directive No. 060 to be cancelled – V.A. Toponogov's bonus to be two weeks' pay instead of a month's. Reason: his work turned out to be fundamentally flawed.

   011. A.V. Chaplik to be reprimanded for deceiving Olenichev. Reason: deceiving Olenichev is like fooling a baby.

   012. E.G. Batyev and G.I. Surdutovich to be reprimanded for playing an open debute behind the locked door. Reason: the ordinary people will never understand."

Yuri Borisovich found his administrative duties irksome and he always regarded himself as a bad director ("I spend my time worrying about toilets instead of about physics"). He dreamed of "worming his way out" and in the end he did so. In this connection he always remembered his meeting with Viktor Weisskopf, who came to Novosibirsk in 1968. They sat and talked for a long time, recalling their youthful days in Göttingen and taking turns to tell each other what had happened to them since they had parted so many years before. There was a lot to tell. At one point, while listening to Rumer's humorous account of the way he had hardly managed to "worm his way out of being director", Weisskopf objected: "Yura, I don't believe that you were a bad director. You just lacked the desire to be a good one. I did have the desire and I became a good director." Yuri Borisovich was not to be persuaded:

"I don't know about you, Vikki, but at my age for the extra money one gets for being director I cannot obtain the pleasures that would compensate for the drawbacks of the job."

From 1967 until his death Yuri Borisovich worked in the Institute of Nuclear Physics. Here he guided yet another generation of students, another band of "youngsters". These "youngsters" now have their own schools, their own institutes, laboratories and departments. "My student Konopelchenko, who outgrew me ages ago, has still not got his own school. Never mind, he has all that to come."

In 1987 Springer Verlag published B.G. Konopelchenko's monograph "Nonlinear Integrable Equations" which is dedicated to Yuri Borisovich. In 1988 World Scientific published a monograph by O.V. Shuryak "The Quantum Chromdynamic Vacuum, Hadrons and Superdense Matter" which bears the brief dedication: "To the memory of my first teacher, Yu.B. Rumer".

On 28 April 1981 Yuri Borisovich Rumer celebrated his 80th birthday. A vast number of congratulatory letters and telegrams were delivered both to Rumer's home and to the Institute of Nuclear Physics. There were telegrams both short and long – three lines, two pages, even ten pages:

"Dear Yuri Borisovich!

Accept these birthday greetings from one of your students. I remember with gratitude my first years studying theoretical physics under your supervision.

M.A. Markov, Academician."

"Dear Yuri Borisovich,

We remember and treasure our friendship with you, our teacher. You taught us how to play with benzole rings, listen to fading sounds and get wet under cosmic showers. You infected us with the sublime Ising disease. You both enlightened and educated us. On this glorious occasion we wish you the best of health, long life and more of that unquenchable curiosity which has illuminated all your life. From your friends at the Landau Institute of Theoretical Physics, Chernogolovka."

"On the occasion of your birthday we take this opportunity to send our best wishes to you, our patriarch of theoretical physics and to express our deep respect and friendly feelings. Sincerely yours, Lifshitz, Andreyev, Dzyaloshinsky, Pitayevskii."

"Your life has been an inspiration. I am happy to have worked and studied with you..."

"Greetings to your students and friends and to you yourself on your birthday. "Nauka" Publishers are preparing a book: "From Three to Five – Studies in Mechanics." The first part was written in England, the second in Switzerland. The plan is to do the third part in Novosibirsk ..."

"Dear Yu.B.

A man of action is never short of time – Siberian folk saying. Keep it up.

Yours, Sasha Dykhne."

"It has always been and still is a joy to work with you and to feel your benign influence on the development of physics and on the education of the physicists ..."

And so on – hundreds of "Dear Yuri Borisovich" messages.

284 Chapter 16. On Pearl Street

Yuri Borisovich's birthday was a day of festivity in the Institute of Nuclear Physics. When we were preparing the invitation cards for the celebratory meeting of the Academic Council the darkroom staff laboured into the night, rejecting their own work and making fresh starts several times until they achieved a result which they thought worthy of the occasion. However, the theoreticians were not happy even then: "Too dark," they said. There was a day left before the invitations had to be sent out – plenty of time for a flat-out effort. Then somebody remembered that most of the darkroom staff were due to work at a vegetable distribution centre that day. The theoreticians willingly volunteered to go and sort through the potatoes and beets and so give the technicians time to do what both groups now agreed had to be done. And it was done.

The great conference hall of the Institute, where the Academic Council was to meet to honour Yuri Borisovich, could not accommodate all who wished to attend. It was even more difficult to find time for all those who wished to speak. In accordance with an unwritten law there was not a single solemn speech on this solemn occasion. And although, as some Novosibirsk University students sang:

"String us together, the four of us here, And we still don't quite stretch to the 80th year" nobody gave any thought to the age of the person who was being honoured – they made jokes about him as if he were their own contemporary.

The "Quantum" students' club started their presentation with a song. An intrepid student accompanied himself on the guitar and sang an irreverent song about Yuri Borisovich in the Saxon dialect to the tune "Mack the Knife" from the "Threepenny Opera".

Then there was a ballet performance – a real one. The day before the lecturers had asked whether they should invite two leading ballerinas from the Novosibirsk Theatre of Opera and Ballet to come and help out. These two dancers were on friendly terms with the INP staff. The students had declined, saying: "We'll manage." And manage they did. The libretto of the ballet "Stages on a Great Journey" was read aloud by the narrator Grisha Falkovich, a young man of imposing appearance (he could have represented at the same time Napoleon in profile and Mother Courage when facing the audience). To the sound of his well-trained voice with its slightly French-sounding "rr's" four no less talented artists danced their hearts out while their performance was recorded on film:

Act One

Göttingen. A young physicist is working on pentoptics. He is visited by Inspiration and Burst of Creativity. He introduces Einstein to pentoptics. All rejoice.

The young physicist – Kobtsev
Inspiration – Azizov
Burst of Creativity – Naumochkin
Einstein – Kukharuk

Act Two

Russia. Unexpectedly, pentoptics turns out to be vital to the study of flutter in aircraft wings. Shutting themselves off from the mundane world, scientists and designers are about to create an Aeroplane. The plane takes shape, not day by day but night by night. Behold, the wing, complete with Flutter, is firmly attached to the Aeroplane.

The Physicist – Kobtsev
The Designer – Azizov
The Aeroplane – Naumochkin
Flutter – Kukharuk

Act Three

Everything is ready for take-off. The Physicist throws a Switch. The Aeroplane follows a Parabolic Trajectory and takes the Physicist to Siberia.
The Physicist – Kobtsev
The Aeroplane – Naumochkin
The Trajectory – Azizov
The Switch – Kukharuk

Word of this irreverent spectacle was to incense the local citizens, who were stirred up further by the newshounds of the Novosibirsk Chronicle. True, no one actually saw the film. Some said it had been "promoted to the reserve" (i.e. put on a shelf to gather dust); others – that by mistake the negative had been placed in the fixer before the developer. This is not important. The main thing is that the journalists were not yet troubling anyone at that moment and Yuri Borisovich, who was as old as the century and as the quantum, felt as young as the rest of that assembly (the average age of which was probably less than 40).

The day before his birthday Yuri Borisovich had received, among the multitude of greetings, a letter from Viktor Weisskopf:

"My dear friend!

I was amazed when I realised that you were about to attain the Biblical age of fourscore years. Time passes quickly but I remember our time in Göttingen as if it were yesterday. Indeed, it is 50 years this year since I was awarded my doctor's degree in Göttingen and so also 50 years since we became friends.

I want to tell you that your friendship has meant a lot to me throughout my life, even though we did not see each other for so long. One way or another your experience has epitomised the tragic era in which we have lived and I have always admired your fortitude. You never lost your all-encompassing curiosity or that incredible capacity to enjoy life which is has remained characteristic of you even after the most dreadful sufferings.

I remember clearly not only our Göttingen days but also our subsequent meetings in Moscow and Novosibirsk even though many years have passed since those meetings, too.

You, of course, have grown older and the colour of your hair has, no doubt, changed but just as I did whenever we met, I still see you as the young Rumer of the Göttingen days.

Unfortunately, my trips to Russia have become less frequent. Health problems have forced me to give up travelling.

Nevertheless, I hope for and await another opportunity for us to meet and discuss all the problems of this disorderly world.

Allow me to wish you and your wife long years of life, health and joy.

Your old friend,
Vikki Weisskopf."

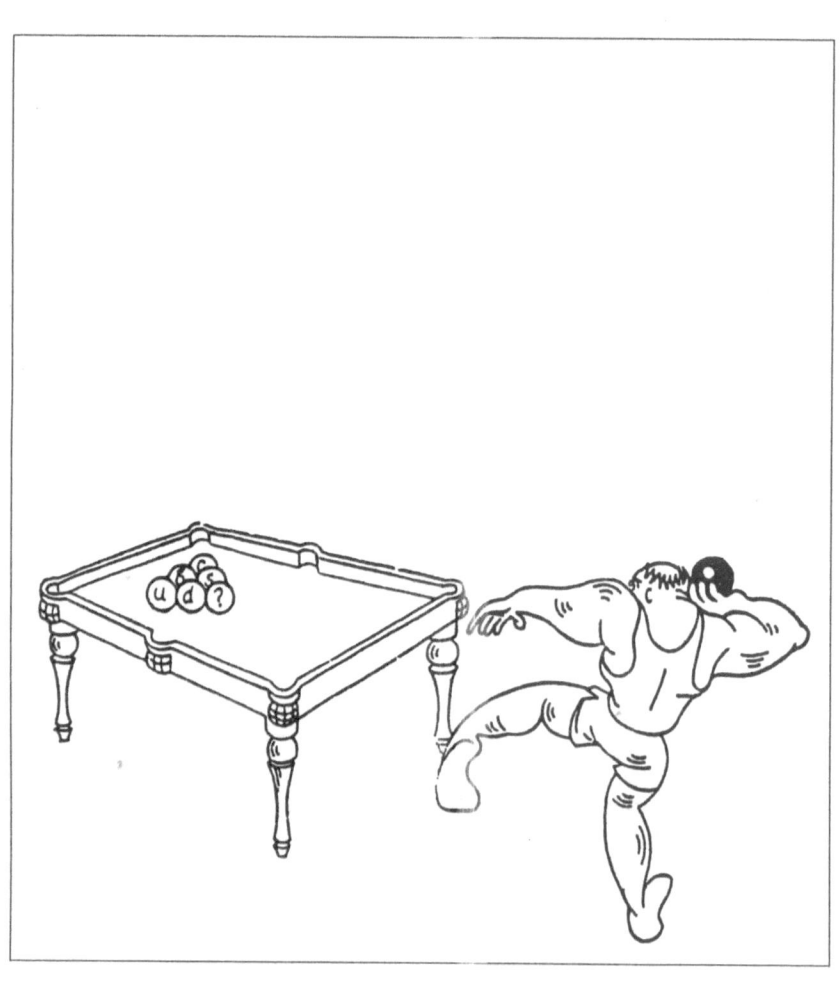

Chapter 17

In the Beginning Was Mechanics

Quantum mechanics, the nuclear physics which came after it and the physics of elementary particles all used to be referred to as "the youngsters' physics". That generation of youngsters whose birth coincided with the birth of the 20th century and the birth of the quantum theory is now passing, or has passed, away. They leave behind them the science that they created. This is the science which has shaped the world we live in and which has led to all the fundamental discoveries of 20th-century physics, chemistry and biology (even suggesting the reason why the primeval symmetry of sugars and acids was shattered, leading to the appearance of life on Earth).

Whatever marvel of modern technology we care to name, its origin can be traced back either directly or indirectly to the discovery of quantum mechanics. This applies as much to the purification of spring water and the protection of grain from weevil damage as to surgical lasers and the highly complex devices which control thermonuclear reactions. Nevertheless, out of all that enormous inheritance which the quantum generation has passed on to us their most important gifts are the new generation that they taught, the schools they founded and the schools which are being founded today.

What, then, is physics today? Of course, this is a question which cannot be answered in just a few words. It is equally impossible to give a short answer to the question "What are physicists working on?" except perhaps an unhelpful one – "Physics is now a mighty industry."

Let us ask a more restricted question: "What is happening at the cutting edge of pure science?"

One of the basic concerns of physics is the nature of matter – the nature of the stuff of which you and I, the Earth and the Universe are made. The ancients, too, pondered as to what matter is made of and came up with an answer – matter is composed of atoms. The word atom means "that which cannot be cut or divided". In other words, matter consists of indivisible particles.

If we accept this view today then we face another question – the answer to which is unobtainable without a colossal financial outlay and a daunting research programme – i.e. "Which particles are indivisible?"

That which we still call the atom can, alas, be split. So can the very nucleus of the atom. Even the neutron, which is one component of the nucleus, can be divided further. However the word "atom", the meaning of which is now clearly at odds with reality, is still used. It was once proposed that the "atom" should be renamed the "tom" (i.e. "that which can be cut or divided") and that even the atomic bomb should be renamed the tomic bomb. However, the idea did not catch on – linguistic inertia has preserved the inexactitude.

Meanwhile we still have to determine which particles or components are truly indivisible, truly "elementary". This question still remains but is now being approached on a completely different level.

In order to convey a proper idea of the level now reached it would be necessary to give a full account of the development of nuclear physics and of the physics of elementary particles and of high-energy physics over the sixty years which separate us from that triumphant time when, thanks to the discovery of quantum mechanics, the general shape of physics was so amazingly simplified. That was the moment when all the hundreds of chemical elements listed on Mendeleyev's table, with all their multitude of different properties, were reduced to just three particles – the proton, the neutron and the electron. It would be necessary to show how this simple picture, which presented itself to the physicists of the 1930s, gradually gave way to a new and catastrophic complexity. By the 1960s the number of recognised elementary particles had risen to over one hundred and now the number is over 350 and still growing.

All elementary particles known today are classified as either hadrons or leptons. The photon, the light particle, has a special status. The class of leptons consists of only six particles (and their corresponding six anti-particles) which are grouped as three pairs: the electron and the electron neutrino; the muon and the muon neutrino; the tau-lepton and its neutrino.

All the remaining particles are classed as hadrons and are divided into baryons and mesons. Baryons are heavy particles which have a mass not less than that of a proton. They also have a spin of one half.

Mesons take a little longer to describe. "Mesos" in Greek means "middle" or "intermediate". The mass of the first mesons to be discovered was mid-way between that of the proton and that of the electron. Later, heavy mesons were discovered but the name was retained. They have whole-number or zero spin.

In addition a great number of excited states of hadrons were discovered. These are the so-called resonances, which have an infinitesimally short life-time of the order of 10^{-23} seconds.

There are, therefore, over 300 particles, each one with its own mass, life-time and characteristic behaviour. Moreover, every one of these particles, whatever its allotted life-time – from 10^{32} years (in the case of the proton) to 10^{-23} seconds – retains its own individuality. None of these particles can be "smashed to pieces". When they collide they simply disappear and other particles are born.

In order to explain the widely differing properties of these particles, in order to understand the dynamics of the various processes in which they participate, it became essential to systematise them by reference to their internal structure. That is, it became necessary to recognise that the elementary particles were not elementary at all but themselves had components.

In 1964 Gell-Mann and Zweig put forward the hypothesis that all hadrons consist of quarks: baryons each consisting of three quarks, mesons of a quark and an anti-quark. The quark idea proved very successful and the situation soon resembled that of the thirties, when the properties of all the elements on Mendeleyev's table were successfully explained in terms of three particles.

At first there were three quarks, too. "Three quarks for Master Mark" is a phrase from James Joyce's novel "Finnegan's Wake". That was where Gell-Mann borrowed the term "quark". What "quark" means nobody knows and nobody can ever know because it is a word that Joyce invented himself. There are those who claim that it is some kind of curd or clay but that is irrelevant. What matters is that quarks are now considered to be the fundamental particles. Their charges, masses and spins have been measured in experiments.

Experiments have also confirmed the existence of the fourth quark, which was predicted by the theory. Further experiments have unveiled a fifth quark and the theoreticians think that there must be a sixth quark – but this is not yet certain. Despite the doubt surrounding it, the sixth quark has been given the most unambiguous of the quark titles: it is called the truth quark. The six quarks, just like the leptons, are grouped in three pairs. The names which the quarks have received, although they do have some kind of history, are mainly arbitrary creations. The pairs are as follows: the up or u (+ 2/3) and the down or d (– 1/3); the charm or c (+ 2/3) and the strange or s (–1/3); the truth or t (+ 2/3) and the beauty or b (–1.3).

The quarks have spin of a half and carry a fractional electrical charge. This charge is given above in brackets after the name of each quark. All the hadrons are built up out of these six quarks. For example, the neutron consists of one u-quark and two d-quarks and is described as udd. The proton is duu.

The charges and masses and all the other quantum characteristics of the hadrons which are predicted by the quark hypothesis are confirmed by experimental results. Such a theoretical structure, however, runs into one big problem. The problem lies in the fact that Pauli's principle applies to all particles with spin of a half. Pauli's principle states that such particles – including the electron, for which it was first stated – cannot be together in one and the same state. So – two d-quarks, for example, making up a neutron must be distinguished one from the other by some property of which we are ignorant. (Let us note that not only Pauli's principle is involved here. There are other serious reasons for the postulation of states relating to a new property.)

It turned out that just three distinct states relating to the new property were enough to explain all the variations in the elementary particles. The new property was dubbed "colour". So – as well as an electrical charge, quarks also carried a "colour" charge. By analogy with the three primary colours of the visible spectrum these three states were named "red", "green" and "blue".

There were now, therefore, not six quarks but eighteen: each of the basic six could occur in red, green or blue varieties. The colour charges combine in hadrons in such a way that the colour charge of the hadron itself is always neutral – i.e. white. This is a very superficial account; in actual fact everything is much more complicated. However, it is generally accepted today that matter is composed of 24 fundamental particles: six leptons and eighteen quarks.

In Moscow there is a wonderful group of young artists. Each member has his or her individual style and subject matter. What unites them is their poster. A poster is a very special challenge for any artist; it requires a very particular kind of talent. A poster is a shout – and how do you paint a shout so that it is compelling rather than irritating? These artists have a poster with an ecological theme. It shows a tiger's head. The tiger is looking straight at the viewer. Every detail, every tiny hair, is painted with such precision that you could mistake the picture for a colour photograph. At the bottom in large letters are the words: *"There are 24 Tigers left in our country."*

Where tigers are concerned 24 is a worryingly small number. Where fundamental particles are concerned 24 is a lot. Such a large number of bricks of which all things, both living and non-living, are supposed to be built up naturally creates unease: surely nature cannot be so complicated! What if these 24 particles have an internal structure and consist of some kind of sub-particles of which in turn there are also too many and which also turn out to have components?

It is just like the soul of Koshchei the Deathless in the Russian fairy tale: at the end of the world is the mighty ocean, in the mighty ocean is an island, on that island there stands an oak, under that oak a chest lies buried, in that chest there is a hare, in the hare a duck, in the duck an egg, in the egg the yolk and in the yolk the soul of Koshchei the Deathless. What it looks like even Koshchei himself does not know; he only knows that his vital strength is held therein and if anyone should find his soul then Koshchei will be no more.

We have to consider the possibility that there are, in fact, no indivisible particles. It could well be that those two or three particles of which we would like everything to be composed simply do not exist. Perhaps there are many particles and they are all divisible? Perhaps there are a lot of particles but a small number of forces acting between them? Here we are looking at the end of the road – a theory which would unite all the elementary particles with all the forces which exist in nature. With such a theory we could then finally answer our question – what is matter? What are we and the Earth and the whole Universe actually made of?

Some might give a simple answer: in nature there is One Force. It is One and is responsible for All: for the solar system; for the fact that the stars shine; for the expansion of the Universe which is scattering galaxies in different directions at an unimaginable speed; for the way a sunflower turns to face the sun; for chemical reactions; for the fact that after the first difficult year of development a child starts to speak.

However, the idea "One Force" does not help to expand our knowledge on the scientific level. What we are really searching for is a single theory which could describe all the possible kinds of interaction which exist in nature – a single theory from which we could extrapolate the actual laws which govern all the vast variety of processes, whether in the realm of elementary particles or in simple experiments in a school laboratory.

The most important feature of any process is the energy level at which it takes place. We know that in their normal state most substances are solid. When a solid body is heated (that is, when a certain amount of energy is expended) it can be melted. With the input of more energy the liquid so obtained can be brought to boiling point and transformed into a gas. When a substance reaches the gaseous state its molecules are separated from each other. By raising the temperature yet again (expending more energy) it is possible to split the molecules into their constituent atoms. The next stage is to ionise the atoms – to remove electrons. In order to ionise, for example, an atom of hydrogen – i.e. in this case to split the single electron from the single proton – 13.5 eV of energy are required. The temperature corresponding to this energy is of the order of 150 000 degrees.

Extremely high temperatures are involved in any research into the nucleus or elementary particles so the appropriate unit of measurement is the GeV. One GeV is equal to 10^9 eV. On a temperature scale this would correspond to a 14-digit number. Depending on the energy and the state of the substance we are dealing with we shall observe different properties of matter and the action of different natural forces.

All the forces acting in nature can now be reduced to four (in actual fact, to three – we shall return to this point) forces which are regarded as fundamental. These are: gravity – the most universal of the forces – which acts between all components of matter; electromagnetism, which acts between charges – like and unlike, static and dynamic; the strong nuclear force, which acts inside the nucleus and brings about the cohesion of the charged protons with the uncharged neutrons; the weak force, which is responsible for radioactive transformations in general and, in particular, for the fact that our Sun shines. The laws governing the action of these forces are different in each case and each has its own "sphere of influence".

Newton's laws are a beautiful description of the force of gravity. Gravitational attraction weakens slowly as distance increases, obeying the law of inverse squares. It acts over great distances, its strength directly proportional to the masses of the interacting bodies. The constant of interaction (the coefficient of proportionality) is a very small value so that gravitational pull is effective only in the case of extremely massive bodies. The gravitational interaction of two billiard balls, for example, is negligible, whereas their interaction with the Earth we can see with our own eyes. Another illustration: if an atom of hydrogen were held together only by the gravitational attraction between the electron and the proton, its diameter would be equal to the diameter of all that portion of the Universe that we can see. In other words, electrical interaction, which keeps the electron near to its proton, is 10^{36} times stronger than gravitational attraction. In the realm of elementary particles, then, gravity can be ignored.

All this is satisfactory as long as we are dealing with familiar objects. However, a full description of gravity required the genius of Einstein. That is not all. It is precisely gravity, so lucidly explained by Newton and fully described by Einstein in the general theory of relativity, which has proved most awkward to accommodate among the latest developments.

Electromagnetic interaction is not so universal as gravity but it is to electromagnetism that we are indebted for everything in our environment, both man-made and natural: the electric lamp, radio, surface tension on liquids, the elasticity of solid bodies, chemical reactions, ordinary friction – even the cohesion of every single atom. Governing all these widely differing phenomena is just one code of laws – Maxwell's electrodynamics. At the point where quantum effects become apparent, quantum electrodynamics takes over, following the principles first clarified by Dirac.

On the basis of his famous equation – the one which summarises "almost the whole of physics" – Dirac predicted the discovery of the first anti-particle. At the heart of Dirac's theory was the unification of particles and forces in one phenomenon – the quantum field. Quantum electrodynamics is the theory of quantum fields. These fields exert their influence with the help of the field quanta – the intermediaries in electromagnetic interaction – that is, photons. One charge learns of the existence of another charge from "messages" which arrive in the form of photons.

Quantum electrodynamics is the most precise of all theories. On it are modelled the theories relating to two other types of interaction – the strong nuclear and the weak nuclear. The quanta of the strong nuclear field – the intermediaries in strong interaction – are the gluons, whose name is derived from the English word "glue". Just as photons bring about the interaction between electrical charges, the gluons bring about the interaction between the colour charges of quarks. Gluons, just like quarks, are coloured. The nature of the strong quantum field is such that, as the distance between the quarks increases, the intensity of their interaction at first slackens then rises to such a level that at distances of the same order as the diameter of a nucleus (10^{-13} cm) it is impossible to pull the quarks apart – the gluons hold them fast. No free quarks have been detected in the course of any experiment so far carried out.

Strong interactions are described in terms of chromodynamics – that is, quantum colour dynamics. Perhaps we have now reached the point at which we can say: "In the Beginning was Mechanics". The words of the great Newton spring to mind:

"I know not what I may appear to the world but to myself I seem to have been only like a boy playing on the sea-shore, and diverting myself in now and then finding a smoother pebble or a prettier shell than ordinary, whilst the great ocean of truth lay all undiscovered before me."

The aim of physics has always been to move towards this great ocean of truth.

The weak interaction, which is responsible for radioactive transformations, is achieved by the exchange of intermediary vector bosons. Unlike photons and gluons, which have no rest mass, the quanta of the weak field are massive particles. Their mass is 100 times greater than that of the proton and is comparable to the mass of a strontium nucleus. The effective radius of the weak interaction is the smallest of all – 10^{-16} cm. The quanta of the weak field form a triplet with the following designations: W_+, W_-, Z_0. The first two carry electrical charges, the third is electrically neutral.

Let us go back into history for a moment. As we have seen, Becquerel discovered the radioactivity of uranium salts at the end of the 19th century. Then Rutherford and the Curies ascertained that there are three kinds of radioactive radiation: alpha rays (helium nuclei), gamma rays (a stream of photons) and beta rays (a stream of electrons). The most awkward of these were the beta rays. When a radioactive source was placed in a magnetic field the alpha and gamma rays behaved as they were expected to – the alpha rays bent always to the same angle and the photons were not affected by the magnetic field. The beta-rays, however, would be deflected at all kinds of angles in precisely similar magnetic fields – that is, on each occasion they had a different energy, instead of the proper value required for the observance of the law of the conservation of energy.

Beta rays seemed to be breaking the most fundamental law of nature. The riddle of the beta rays remained unsolved right up until the beginning of the 1930s. In 1931 Pauli came up with a bold hypothesis, suggesting that beta rays consist not only of electrons but also of undetectable particles which have no charge and almost no mass. The energy of these mysterious particles must adjust itself in relation to the energy of the electrons in such a way that the total energy always remains the same and does not break the law of the conservation of energy.

Pauli's wonder particles found no support whatever and even Pauli himself did not defend his hypothesis very vigorously. Only after the discovery of the neutron did Pauli again return to this idea, suggesting that the source of beta rays is the decay of the neutrons of the nucleus into protons, electrons and these same mystery particles without charge or mass. He presented this hypothesis at the Solvay congress of 1933.

Soon afterwards Enrico Fermi constructed his phenomenological theory of beta-decay. He gave Pauli's particle a truly Italian name – the neutrino. One should not judge the great but it has always seemed to me that Fermi was a little unfair – he could have called it the paulino. It seemed that the mystery of beta-radiation had been solved. Neither Fermi nor anyone else supposed that the enigma would remain for many years yet or that, even so, it would not stand in the way of the development of atomic reactors and bombs in which beta-radiation plays a vital role.

The complete formulation of the theory of weak interactions was achieved in 1971. Most significantly the theory not only accounted for weak interaction and posited the existence of the massive particles (even indicating their mass) which act as intermediaries in this interaction – it also contained within itself the theory of electromagnetic interaction.

Exchange of W-bosons, as has been noted, occurs at distances of the order of 10^{-16} cm. At such distances the mass of the W-bosons is negligible but their potential for interaction is similar to the Coulomb potential – that is, the potential of the interaction of the electrical charges.

The intermediate vector bosons and the photons have now been shown to be equal transmitters of electro-weak interaction. If this theory were to be substantiated then it would be possible to celebrate the achievement of an important step towards the unified description of all the various forces of nature. In 1979 Glashow, Weinberg and Salam shared a Nobel prize for their work on the electro-weak theory. And, although indirect (but very important!) evidence in support of the electro-weak theory was already available people did ask a teasing question: would the prize-winners return the money if they failed to find the vector bosons?

In an Abstract of a short article that was published in 1983 it is stated that: "Not only the geometry but also the number of events are found to be in accordance with what is predicted in the process $\bar{p} + p \rightarrow W^{\pm} +$ something, with $W \rightarrow e + \nu$, where $W^{\pm}$ is the intermediate vector boson postulated by the unified electro-weak theory." This article is the work of 59 authors and over each of their names a small letter is printed to indicate which of the world's laboratories each is attached to. The experiment was carried out at CERN on the super proton synchrotron (SPS).

In the summer of that same year, 1983, Z^0-bosons were also obtained on this same synchrotron. This time the number of authors reporting the event rose to 138. Intermediate vector bosons had been discovered. Their masses coincided exactly with those predicted by the electro-weak theory. This was a real triumph. The number of fundamental interactions had been reduced to three: gravity, strong and electro-weak.

What next? The next thing is energy. If the unification of the electromagnetic and weak interactions takes place at an energy of the order of 100 GeV and this was the energy reached when the W-bosons were discovered, then the energy at which we can expect the unification of the strong and electro-weak interactions is estimated at 10^{15} GeV. This is a colossal amount of energy. The above-mentioned SPS accelerator, on which the unification of electromagnetic and weak interactions was confirmed, has a diameter of 2.2 km, a circumference of 7 km and develops magnetic fields of tens of thousands of Gauss (the magnetic field of the Earth is 1 Gauss). If we could build an accelerator around the whole equator, create colossal magnetic fields and avoid all unwanted effects, then we could achieve an energy level of the order of 10^7 GeV. Not enough. In order to make up the missing eight orders of magnitude we would need an accelerator with a diameter hundreds of times greater than the diameter of the solar system.

As for the unification of gravitational and strong/electro-weak interactions, it should take place at an energy of 10^{19} GeV. This is the so-called Planck energy. If it ever was reached then it can only have been at some time during the first 10^{-43} seconds after the Big Bang which, according to the present theory, gave birth to the Universe. This is the situation facing the poor experimenters.

As for the theoreticians, they have no trouble with such energies – all they need is a pencil and paper. Apart, that is, from the ability to be a good theoretician. Einstein spent the last 35 years of his life on his search for a unified theory of electromagnetism and gravity. Without success. Before him Riemann dreamed of finding the link between "electricity, galvanism, light and gravity". He not only dreamed but actually found this link and considered that he had made good progress in this enquiry. The most striking thing is, and this has already been mentioned, that Riemann not only sought the unity of these forces but also connected them with the curvature of the space in which we live.

Riemann, however, was born too soon. It was Maxwell who took the first real step towards the unification of the forces of nature. In Maxwell's time it was accepted that all natural phenomena could be described by the laws of mechanics. Even Maxwell found it difficult to break through this obstacle but he showed that electricity and magnetism are inherent in nature, that they cannot be explained by the laws of mechanics and that these two apparently independent properties of nature are in fact different manifestations of one and the same phenomenon – electromagnetism.

The next step, as we know, was taken only 120 years afterwards – in our own time, with the creation of the electro-weak theory. In one of his lectures Salam said: "Let me try to show you how ideas about the fundamental forces have changed. I remember a lesson given by my first physics teacher in 1935 when I was at school in my home town of Jangmaghiam in Pakistan. The teacher told us about gravity and Newton's theory. Then he went on to magnetism, magnets being available even in Jangmaghiam. He said that magnetism was also a fundamental force. Then he said that there was another force called electricity but that it "could be found" only in Lahore, the state capital 50 miles to the east of us. As for nuclear forces, they were only to be found in Europe." [52, p. 177]

Now Salam is one of the authors of the electro-weak theory. The next step – the unification of the electro-weak and strong interactions – is one of the main problems facing fundamental physics. The models that are being constructed differ widely. Many of them are very elegant. The problem is discussed by schools, conferences, symposia and seminars. The magic set of initials perhaps most favoured nowadays is GUT – the Grand Unification Theory. GUT must, of course, encompass gravity – but that is where the greatest difficulties lie.

As we know, Einstein completed the theory of gravity, realising Riemann's dream of connecting the forces of nature with the curvature of space. The brilliance of Einstein's theory lies in its identification of the inertial mass of an object with its gravitational charge, which is expressed through the curvature of four-dimensional space-time. Abdus Salam wrote: "The key to Einstein's achievement (in my opinion, the greatest in the history of physics) is his recognition of the fundamental significance of charge in gravitational interaction. I want to emphasise that until we understand the nature of charges in electromagnetic, weak and strong interactions as fully as Einstein understood charge in relation to gravity, there is little hope of success with "complete unification". [52, p. 194]. He also says: "We would like not only to carry on with Einstein's efforts, which he did not carry to fruition, but also to include in our programme the remaining charges (i.e. the charges of weak and strong nuclear interactions)" [52, p. 196] This last sentence is full of significance. "Einstein's efforts" of which Salam speaks represented 35 years of work by the greatest of the physicists who have tried to bring together gravity and electromagnetism by means of the geometrication of the interactions. Einstein tried various approaches. One of them involved rejecting Riemann's dimensional scheme and turning to more general non-Riemannian geometries. Another approach, which became the main one, involved the introduction of a fifth dimension to Riemann's geometry. Kaluza had done the same thing in an article which he published in 1921. Kaluza, in fact, succeeded in constructing a single theory of electromagnetism and gravity in five-dimensional Riemann space.

In 1926, after the discovery of quantum mechanics. O. Klein borrowed Kaluza's idea and developed further a five-dimensional theory of electromagnetism and gravity. We shall say no more of what became of these endeavours. We shall say no more about the work of Einstein himself, about Einstein and Bergman, about Rumer's pentoptics or about the multitude of other five-dimensional projects. Interest in these approaches gradually faded. The results of the work on the nucleus and on elementary particles and their interaction was far more exciting.

Ironically, however, the physicists learned so much from the new developments that the programme that Salam spoke of was put on the agenda for further discussion. This was a programme which aimed not only to continue the work of Einstein – and of Kaluza, Klein, Rumer, Jordan and many others, too – but also to bring the strong and weak interactions into the same framework. Once again the need arose for multi-dimensional generalisations. Once again there was a return to Riemann's dream of linking the forces of nature with the curvature of space. Once again people started quoting the articles of Einstein and Bergman, Kaluza and Klein, Rumer and Jordan – work that goes back 30, 40, 50 and more than 60 years.

Books with bold titles began to appear – such as "Unified Field Theories Using More than Four Dimensions", to which were added the words "Complete with Exact Solutions". No-one could remain indifferent to such a promise – or read it without a smile. Apart from purely mathematical ones, no exact solutions relating to multi-dimensional frameworks are yet to hand. However, there are some intriguing ideas. There are, for example, attempts to unite all the charges – the graviton (the quantum of the gravitational field), the photon, the intermediate vector bosons and gluons – in one charge connected with the curvature of eleven-dimensional space.

We, of course, know that our world has four dimensions – height, length, width and time. What then has happened to the other dimensions, how are they manifested, how are we to imagine eleven-dimensional space if it truly is eleven-dimensional? Here we cannot help noticing the way "the smallest" is now linked to "the largest" – i.e. how closely the physics of elementary particles relates to cosmology.

In the standard cosmological model the biography of our Universe begins with the so-called Planck period, 10^{-43} seconds from the moment of its birth. The earlier moments lie beyond the limits of the theory of gravitation. 10^{-43} seconds after the Big Bang our Universe was a tiny incandescent ball the size of which it is impossible to imagine: the diameter of the ball was equal to the so-called Planck length – i.e. 10^{-23} cm. The temperature of the ball – 10^{32} degrees – is also impossible to imagine. This temperature corresponds to the Planck energy 10^{19} GeV. The density of our Universe at that moment is supposed to have been 10^{90} kg/cm^3.

Up to that point there had been no nuclear or electro-weak charges – there were only the strong effects of quantum gravitation and one charge corresponding to this unimaginable gravitational field. And if there was this charge then possibly it was connected with the curvature of the equally unimaginable eleven-dimensional space.

By the time the Universe was 10^{-35} seconds old it had expanded so much that its temperature had fallen by five orders of magnitude (100 thousand times) and the quarks, which had by that time appeared, had begun to interact. The energy at this moment was of the order of 10^{14} GeV. At this energy intense interaction took place and the synthesis of quarks began. Hadrons were formed but bosons and electrons were still far off in the future.

As the expansion of the Universe proceeded (while we are in the hadron era we have to remember that the concept "proceed" is used of events taking thousandths of a second – the time scale is still very compressed) weak interactions began and with them radioactive disintegrations (for example, the disintegration of free neutrons into protons, electrons and neutrinos) and the lepton era dawned.

As the age of the Universe approached one second, typical energies dropped to 10^{-3} GeV and the creation of neutrons slowed down. However their energies were still high enough for them to react with the protons.

By the time the Universe was 100 seconds old energies had fallen to 10^{-4} GeV and the synthesis of nuclei had begun. Thus began the helium era. From that moment on, the time scale started to stretch out considerably. The Universe was now a "soup" of helium nuclei, deuterons, free electrons and neutrinos.

The soup evolved slowly. It was not until 10^6 years later, by which time the Universe had expanded so much that typical energies had fallen to 0.1 eV (a mere 1000 °C) that the formation of atoms began and matter became distinct from radiation, from photons. After this began a slow process which much later, 10^{10} years after the Big Bang, led to the formation of stars and galaxies, of our solar system and our Earth.

For the moment the Big Bang model is working well and there is cosmological evidence to support it. There are difficulties but there are also incontrovertible facts. Let us return now, however, to that moment 10^{-43} seconds after the Big Bang. Straight after this instant, as soon as the Universe exceeded the Planck size of 10^{-33} cm, as soon as quarks appeared and started to interact, eleven-dimensional space with its single charge of strong quantum gravitation changed: seven dimensions of space were conflated and twisted themselves into a ring with a radius of 10^{-33} cm. From that moment on and to this day the Universe has seemed to be four-dimensional. We must remember that it only appears to be four-dimensional and that in reality it is eleven-dimensional – but only as long as its radius is 10^{-33} cm! This is roughly the same as going down with flu after catching a virus. We cannot see the virus with the naked eye but we can see it with a microscope and the electron microscope reveals that the virus is a whole miniature world.

Accelerators are the microscopes that we use to study elementary particles. When W-bosons were discovered we were observing events that took place within a space 10^{-16} cm across (the effective radius of weak interaction). That is the scale on which we are now able to "see". As we have seen, it took 100 GeV of energy to get us to that level. In order to make out what is happening in spaces measuring 10^{-33} cm across we shall need 10^{19} GeV – Planck energy.

Speaking about the eleven-dimension theory in one of his lectures Salam gave special emphasis to these words: "If this theory is correct then, possibly, we are very close to the full and final unification of all the forces ... moreover the fundamental charges are accounted for as manifestations of the hidden dimensions of space!" [52, p. 201]

It is difficult to predict the direction that the experimenters will take in their efforts to confirm or disprove the Grand Unification Theories that are now being formulated. There is plenty of work for both theoreticians and experimenters to do. At this point I would like to return to the words of Bishop Sprat, which were quoted at the beginning of the book, concerning "the third kind of new philosophers" who were recognised as true scholars only in the 17th century. What the Bishop said is still appropriate: "Much has already been done ... One may have doubt concerning only the attitude of future ages. And even then we can safely promise that they will not long be deprived of a whole galaxy of enquiring minds, for before them lies such a clearly marked path. They have only to taste these first fruits to be inspired by this example."

Prophetic words, indeed. The example did truly prove to be inspiring but the "clearly marked path" had to take a radically different route in order to lead to further successes. We may think that our path today is clearly marked but we, too, are unable to place our own work in a long-term perspective. The right to judge us, to see the links between the achievements of our age and the discoveries which are yet to come, belongs to future generations.

Since we have mentioned other generations let us fantasize a little. Let us go back in imagination, not as far as Bishop Sprat's contemporaries but to the people who lived 100–150 years ago. If we could raise them from the dead and show them how we live, everything that they saw would seem to them to be miraculous. Moreover, we need only show them our most mundane technology, items in everyday use such as telephones, television sets, jet aircraft, children's electronics kits. They would be amazed to see how a ten-year-old boy could, in half an hour and not even using a soldering iron, screw transistors and resistors into the right places to construct a simple device which would bring a clear signal from the silent ether into his earphones. By turning a knob he can pluck either a fairy tale or a song from the void. Then let us ask ourselves what we would find if we could return to life 100 – 150 years from now – what wonders would we see which future generations will take for granted? I have asked people of various ages and professions to consider this question. Probably the most accurate answers were given by scientists, whose experience has taught them to be cautious. Their various answers can be summed up as follows: it is impossible to answer this question. You cannot predict a wonder. That which is predictable now will not seem like a miracle then.

This approach may seem pedantic but it does show the logic of the scientific mind. Let us remember the story of weak interactions – how the existence of very heavy particles was postulated in order to account for the interactions of the light particles. This, in the words of John Adams, the supervisor of the super proton synchrotron, was: "as if you struck two pocket watches sharply together and instead of getting a pile of gear wheels you end up with a grandfather clock." [53, p. 240]

The "grandfather clock" was in fact discovered. From the scientific point of view this was not a miracle but the result of the everyday work of physicists. As for miracles, children and fantasy writers invent them without difficulty. Children imagine, for example, that in 100 years' time learning lessons will be a thing of the past. You swallow an English language tablet and you get top marks in English; you swallow a science tablet and you know all about physics.

"My grandmother is ill at the moment", said one girl, "I love her very much. Perhaps in a hundred years' time there will be a device like a glass sphere. The patient will get into it, the instruments will be switched on and signals will appear on screens. Then a computer will take these signals and relay them to a more complex device which will go into action and remove the cause of the illness."

A twelve-year-old boy thought for a while about the question and then launched into a detailed argument. He thought that miracles 100 years from now are not necessarily to be expected. Human society, he pointed out, does not simply develop at a constant rate. When people lacked a lot of things they invented a lot of things. Now, however, so much has been created that, at least where technology is concerned, the future will be one of "developing and perfecting". For example, perhaps in 100 years aeroplanes will no longer need wings – or pilots, for that matter. Everything will be automated. Passengers will get into a wingless machine, a controller will press some buttons and the aircraft will land in the right place at the right time, regardless of weather or season. That will not be a miracle, just an everyday thing. In fact everything will be perfected and will change roughly in the same way as automobiles have developed from the first comical "horseless carriages" to the modern car. Their appearance may be very different but their engines work on the same principles, even if the modern engine is placed transversely. So there.

Just the answer given by this boy, who has written computer programs since he was nine years old, would be a wonder for our forebears. At the end of his serious reflections he suddenly laughed out loud and said: "Yesterday we finally got our new program working properly. You know, there was once a physicist named Rutherford. He had a lot of problems but he bombarded his atoms and found out that inside the atom there is a nucleus. On our machine it's no problem at all. So you see, we got the whole program sorted out and there was still half an hour of machine time left. Our teacher Gennadi Anatolyevich – whom we call "Greeny" – told us to run through the program again just to make sure. We knew, though, that this was a waste of time so we programmed the computer to write *Greeny* on the screen 1000 times. Gennadi Anatolyevich looked at the monitor and each time his name appeared he said: 'I have to put up with all this, I have to put up with all this'. Perhaps he would have said it all the 1000 times but he was called to the telephone."

For this boy electronic watches and diskettes are everyday objects just as clockwork toy motorcycles and paper Christmas tree decorations were familiar to our post-war generation. He wants to be a biophysicist when he grows up! "Living nature is more interesting than crystals but first I need to study physics and absorb all I can from it. People need to study themselves."

Here I cannot help recalling another answer to the question about the wonders of the future. It is the answer given by an old Hindu, a man wrinkled by years and cares, who had spent his life delivering babies for the women of his tribe and had never left his little village near Jaipur. He played with his beads and took his time before answering. What he said went something like this. At the same moment in prehistory human beings began to make tools and hunting weapons, began to draw and sculpt, to compose songs and legends, to build homes. Over time everything changed – at least, tools and houses changed, but the stories stayed the same. This is simply because in stories people dream always of the same things; happiness, love, the victory of good over evil. We sometimes use the phrase "just like a fairy tale", usually about something pleasant which we feel has happened of its own accord, by some kind of magic. But really in a fairy tale nothing happens by itself. In order to triumph, the force of good has to pass through many hard tests. There is evil everywhere. It is strong and cunning and no laws can bind it. If we do not help the good, then it may not conquer the bad.

Fairy tales are full of flying carpets, obedient djinns, magic wands which can dry up the sea or fill it to overflowing, magic mirrors which bring images from afar, table-cloths that provide food, even living water. People dreamed of travelling fast, of crossing land and sea, of seeing and hearing at a distance. All this has been achieved in reality.

"Probably food-producing table-cloths are not too far off, either. I hear that in your country you have been making caviar from paraffin for a long time. However, this dream – of giving every human being on Earth enough to eat – has not yet come true," said the old man very seriously, "and people will continue to strive to achieve this miracle. Living water, too, wonder of wonders – that will become the main aim of people in the future. People will never find living water, however, until they make pure the heaven over the earth, until evil intent is purged from their hearts, until kindness is what they value most. This is what people will have to understand."

References

1. The Born-Einstein Letters. Correspondence between Albert Einstein and Max and Hedwig Born from 1916 to 1955 with commentaries by Max Born. London: McMillan 1971
2. Einsteinovskii Sbornik, 1972. Moscow: Nauka 1974
3. Bernal, Dzh.: Nauka v istoriyi obshchestva. Izd-vo inostran. lit. 1956
4. Planck, M.: Sbornik k stoletiyu so dnya rozhdeniya Maksa Planka. Moscow: Izd-vo AN SSSRI 1958
5. Dzheff, B.: Maikelson i skorost' sveta. Moscow: Izd-vo inostran. lit. 1963
6. Akhmatova, A.: Poema bez geroya. Stikhotvoreniya i poemy. Sverdlovsk: Sred.-Ural. kn. izd-vo 1987
7. Planck, M.: Predisloviye ko 2-omu izd. Teoriya teplovogo izlucheniya. Leningrad, Moscow: ONTI 1935
8. Born, M: Fizika v zhizni moyego pokoleniya. Moscow: Izd-vo inostran. lit. 1963
9. Zelig, K.: Albert Einstein. Moscow: Atomizdat 1966
10. Klein, B.: V poiskakh. Fizika i kvantovaya teoriya. Moscow: Atomizdat 1971
11. Robertson, P.: The Early Years. The Niels Bohr Institute 1921–30. Universitets-forlaget Copenhagen: Akad. Forlag 1979
12. Bohr, N.: Nature *116* (1923) 845
13. Pauli, W.: Z. Phys. *31* (1925) 373
14. Teoreticheskaya Fizika 20 veka. Moscow: Izd-vo inostran. lit 1962
15. Lyottsi, M.: Istoriya fiziki. Moscow: Mir 1970
16. Hoffman, D.: Erwin Schrödinger. Moscow: Mir 1987
17. Schrödinger, E.: Izbranniye trudy po kvantovoy mekhanikye. Moscow: Nauka 1976
18. Rumer, Yu. B.: Neizvestniye fotografiyi Einsteina. Priroda: 1977 Nr. 9
19. Göttingen. Album Göttingen und Umgegend. Vereinigung Göttinger Papierhändler 1910
20. Rumer, O.: Izbranniye perevody. Moscow: Sov. pisatel' 1959
21. Konstantinov, N.: Ocherki po istoriyi sredney shkoly (Gimnaziyi i realiiiye uchilishcha s kontsa XIX v. do Fevralskoy Revolyutsiyi). Moscow: Uchpedgiz 1947

22. Zarnitsky S.V., Trofimova L.N.: Sovietskoy strany diplomat. Moscow: Politizdat 1968
23. Lusternik, L. A.: Molodost' moskovskoy matematicheskoy shkoly UMN. 1967. T. 22, vyp. 1/2, 4, 1970. T. 25, vyp. 4
24. Aleksandrov, P. S.: Stranitsy avtobiografiyi UMN. 1979. T. 34, vyp. 6; 1980. T. 35, vyp. 3
25. Rid, K.: Hilbert. Moscow: Nauka 1977
26. Gauss, C. F.: Werke. Bd. VIII. Leipzig, 1900. (see also Norden A. P.: Ist.-Mat. Issled. Moscow: Gostekhizdat 1956, vyp. 9)
27. Kagan, V. F.: Ocherki po geometriyi. Moscow: Izvd-vo MGU 1963
28. Laptev, B. L.: Nikolay Ivanovich Lobachevsky. Kazan: Izd-vo Kazan. univ-ta 1976
29. Boljai, Janos.: Appendix. Prilozheniye, coderzhashcheye nauky o prostranstrvye absolyutno istinnuyu. Moscow; Leningrad: Gostekhizdat 1950
30. Livanova, A.: Tri sud'by. Moscow: Znaniye 1975
31. Riemann, B.: Sochinyeniya. Moscow; Leningrad: OGIZ 1948
32. Einstein, A.: Sobr. Nauch. Trudov. T. 4. Moscow: Nauka, 1967
33. Joffe, A. F.: Vstrechi s fizikami. Moscow: Fizmatgiz 1962
34. Rumer, Yu. B.: Stranichki vospominanii o L. D. Landau. Nauka i Zhizn' 1974. Nr. 6
35. Weisskopf, V.: Fizika v 20-om stoletiyi. Moscow: Atomizdat 1977
36. Frenkel', V. Ya.: Paul Ehrenfest. Moscow: Atomizdat 1977
37. Albert Einstein in Berlin 1913–33. Darstellung und Dokumente. Berlin: Akad. Verlag 1979
38. Physics Today 198 1. Vol.4, Nr. 11
39. Heisenberg, W.: Der Teil und das Ganze. Munich: 1969
40. Irving D.: Virusnyi fligel'. Moscow: Atomizdat 1969
41. Goudsmit, S.: Missiya "Alsos". Moscow: Gosatomizdat 1962
42. Journal de Physique, Colloque Nr. 8, 1982
43. Khromov, S. S. (ed.): Istoriya Moskvy. Moscow: Nauka 1980
44. Fowler, R.: Noveyshiye dostizheniya v oblasti izucheniya atomnykh yader. UFN. 1933. T. 13, vyp. 1. s. 37
45. Rezolyutsiya po povody statey "Pravdy" "0 vragakh v Sovietskoy maskye" i "Traditsiyi rabolepiya"; "Izzhit'lyuzinshchinu v nauchnoy sredye". UMN. 1937, vyp. 3
46. Janouch, F.: Lev D. Landau: His life and work. Geneva: 1979
47. Nauchnyi arkhiv SO AN SSSR, Novosibirsk. F. 21, Op. 1
48. Rumer, Yu. B.: Issledovaniya po 5-optikye. Moscow: Gostekhizdat, 1956
49. Kerber, L. L.: A delo shlo k voinye. Izobretatel' i ratsionalizator. 1988. Nr. 3–9
50. Akademiya Nauk SSSR. Sibirskoye otdeleniye. Khronika. Novosibirsk: Nauka 1982
51. Vavilov, S. I.: Isaac Newton. Moscow: Izd-vo AN SSSR 1961

52. Salam, A.: Unifikatsiya sil. Fundamentalnaya struktura materiyi. Moscow: Mir 1984
53. Adams, J.: Instrumenty fiziki elementarnykh chastits. ibidem

Biography

Akhmatova, Anna (Anna Gorenko, 1889–1966)
Russian poet and writer, one of the most popular among Russian "intelligentsia", died in poverty as objector to regime.

Alexandrov, P. S. (7. 5. 1896–17. 12. 1982)
Mathematician, founder of topology. The president of Moscow Mathematical Society from 1932. Member of Soviet Academy of science, of the National Academy of USA, and others. Lobachevsky prize winner.

Archangelsky, A. A. (1892–1978)
Aircraft designer, Hero of Soviet Union. Participated in creating almost all the aircraft in Tupolev's Design Bureau. Three State prizes (1941, 1949, 1952), one Lenin prize (1957).

Bari, Nina (19. 11. 1901–15. 7. 1961)
Mathematician, Professor at Moscow University since 1934. Fundamental results in the theory of functions of real numbers, theory of orthogonal series, and others. Member of French and Polish Mathematical Societies.

Baidukov, G. Ph. (1907)
Air force general, Hero of Soviet Union (1936). Writer. Participated in

nonstop flight together with Chkalov and Belyakov from Moscow to Vancouver via North pole.

Belyakov, A. V. (1897)
Air force general, hero of Soviet Union (1936). Participated in nonstop flight Moscow–North Pole–Vancouver.

Blok, Alexander (1880–1921)
One of the most beloved Russian poets.

Bothe, W. (8. 6. 1891–8. 11. 1957)
German physicist, one of the pioneers of nuclear physics, student of Max Planck. Nobel prize for a new method in analysis of cosmic rays, 1954.

Brik, Lilya (KAGAN)
Well educated beautiful woman. Wife of Osip Brick who played a role in the Moscow literary society in the 1920s, and who was a cousin of Yuri Rumer. Mistress of Vladimir Mayakovsky who devoted to her a poem and several sonnets.

Chaplygin, S. A. (5. 4. 1869–8. 10. 1942)
Professor in applied mathematics, fundamental contributions in theoretical mechanics and hydroaerodynamics, student and co-worker of Zhukovsky. Director "TsAGI" (Central Aero-Hydrodynamic Institute) from 1921. Petersburg gold medal (1900), Zhukovsky prize (1925). In 1942 Soviet Academy of Science established Chaplygin prize. One of the craters on the moon carries Chaplygin's name.

Chicherin, G. V. (1872–1936)
Member of Russian Democratic party (1905), Russian, and then USSR minister of Foreign Affairs (1918–1936). Carried out scientific research on international politics.

Chkalov, V. P. (1904–1938)
Pilot, hero of Soviet Union (1936). In 1936–37 – nonstop flights: Moscow- Udd (Far East) and Moscow–North Pole–Vancouver. Perished in air crash.

Egorov, D. F. (22. 12. 1869–10. 9. 1931)
Mathematician. Main fields of research – differential geometry, theory of integral-differential equations, variational principles, and others. There are several theorems carrying his name.

Eisenstein, S. M. (1898–1948)
Producer, researcher in theory of cinema, professor, writer. Most famous movies: "Battleship Potemkin", "October" (1927), "Alexander Nevsky" (1938), "Ivan the Terrible".

Flyerov, G. N. (2. 3. 1913)
Physicist, member of Soviet Academy of sciences. Worked in Kurchatov's laboratory (1943–1960). Since 1960 the director of Nuclear Reactions Laboratory in United Institute of Nuclear Research in Dubna. State prize (1946). Hero of Soviet Union.

Fock, V. A. (22. 12. 1898–27. 12. 1974)
Physicist, theoretician. One of the creator of modern physics. There are "Fock's Equations", "Focks symmetry conditions", "Fock's representations", "Fock's transformation set, "Hartree-Fock method", "Klein-Fock-Gordon Equation", etc. Academician, Hero of USSR, member of many Academies. A friend of Peter Kapitza.

Franck, James (26. 8. 1882–21. 5. 1964)
Physicist. Professor at Göttingen University and the director of Physical Institute in Göttingen (1920–1933). Discovered experimentally the scattering laws between the atoms and electrons. Nobel prize in 1925. Fundamental results in photo-chemistry and in spectral studies of chemical forces. Participated in American atom bomb project.

Frenkel, Ya. I. (9. 2. 1894–23. 1. 1952)
Physicist, theoretician. One of the founders of modern theoretical physics. Made fundamental contributions to the theory of solid state and fluids, quantum field theory, nuclear physics and physics of elementary particles. In 1928 gave the quantum theory of ferromagnetics. In 1936 worked out the drop model for nuclei. In 1939 predicted the spontaneous fission of nuclei. Wrote more than 20 books. Among them "Statistical Mechanics", "Electrodynamics", "Kinetic Theory of Fluids", and many others.

Gelfand, A. O. (24. 10. 1906–7. 11. 1968)
Mathematician, professor in Moscow State University (1931). Fundamental results in the theory of numbers and theory of complex functions. The founder of the theory of transcendental numbers. Solved Hilbert's 7th problem of transcendental numbers. Member of several academies of sciences and the International Academy of the History of Science.

Glushko, V. P. (1908)
Pioneer of the rocket industry and founder of rocket engine technology in Soviet Union. Constructor of the first electro-thermal rocket engine(1929), and first Soviet liquid rocket engines (1930–31). Academician. Twice Hero of Soviet Union (1956, 1960). Lenin prize (1957), State prize (1967).

Goldbach, Ch. (18. 3. 1690 –1. 12. 1764)
German mathematician and lawyer, member of Petersburg Academy of Sciences. From 1742 worked in Moscow in the Ministry of Foreign Affairs. In a letter to Euler formulated the mathematical problem in the theory of numbers which was not solved until 1930 (when it was solved by Lev Schnirelman) . Achieved fundamental results in solving differential equations of Riccati.

Gromov, M. M. (1899)
Air force general, Hero of Soviet Union (1934). The world record (1934) for distance of nonstop flight (more than 12000 km). Nonstop flight from Moscow to USA via North pole, together with A. Yumaschev and S. Danilin.

Herglotz, G. (2. 2. 1881–23. 3. 1953)
German mathematician and physicist. Worked in Leipzig and Göttingen. Main field of research – theory of functions, differential geometry, the theory of numbers, applications to general and space mechanics. There are the Herglotz equation in differential geometry, the Herglotz theorem and Riesz–Herglotz multiple expansion.

Joffe, A. (29. 10. 1880–14. 10. 1960)
The organizer of physical research in Russia and Soviet Union. Started his own research in 1902 in laboratory of Roentgen (Munich). Made fundamental contributions in solid state physics and in a wide range of experimental physics. State (1942) and Lenin (1961) prizes. Hero of Soviet Union. Member of many Academies of science.

Kamensky, V. V. (1884–1961)
Russian poet. In early days futurist. Later wrote the romantic poems "Sten'ka Rasin" (1912–1920), "Emelyan Pugachev" (1931), "Ivan Boltnikov" (1934). One of the first Russian pilots. Introduced the Russian word for airplane "samolyet".

Keldish, M. V. (10. 2. 1911–24. 6. 1978)
Mathematician. Main theoretician in Soviet Space research programs (1961–1975). Member of many Academies of science. Three times Hero of Soviet Union (1942, 1946, 1971). The president of Soviet Academy of Science (1961–1975). One of the small planets and one of the moon craters carry his name.

Khariton, Yu. B. (27. 2. 1904)
Physicist. Worked in Joffe's Institute (Leningrad) and then in the Cavendish laboratory with Rutherford. In 1939 together with Zeldovich calculated the chain reaction in the uranium fission. Three times Hero of Soviet Union for essential contributions together with Sacharov and Zeldovich to the A- and H-bomb.

Khlebnikov, Velimir (1885–1922)
Very popular Russian poet, futurist. Mathematician, philosopher. Died of starvation.

Kleimenov, I. T. (13. 4. 1898–1938)
The head of research and project in rocket technology. In 1933–1937 the director of "Reactive Institute", where the best scientists and engineers in research of jet theory and technology were gathered. Due to a false denunciation was arrested and shot. One of the craters on the far side of the Moon carries his name.

Krutkov, Yu. A. (29. 5. 1890–12. 9. 1952)
Physicist, theoretician. One of the founders of physics research in Russia and Soviet Union. Fundamental contributions in quantum physics, statistical and solid state physics, mechanics, and others. Member of Academy of USSR (1933). State prize (1952). Excellent lecturer.

Korolyov, S. P. (1907–1966)
Designer of the first space rockets. The founder of practical aeronautics. The head of Soviet space research program. Twice Hero of Soviet Union (1956, 1961). Lenin prize (1957).

Lakhtin, L. K. (1858–1927)
Professor of pure mathematics. Representative of a strong Russian school of mathematics. Founder of Soviet school of applied statistics.

Landau, Edmund (14. 2. 1877–19. 2. 1938)
German mathematician. Major contributions to complex functions and the analytical theory of numbers. There is a Landau theorem on singular points. Wrote several fundamental books on mathematics. One of the Moon craters carries his name.

Langemak, G. E. (1898–1938)
Designer of Soviet gunpowder rockets. Worked on rocket missiles for famous "Katyusha". Following a false denunciation was arrested and shot.

Leipunsky, A. I. (7. 12. 1903–14. 8. 1972)
Physicist. In 1932 together with Valter, Sinelnikov and Latyshev realized lithium fission by artificially accelerated protons. In 1934 found first indirect justification of neutrino hypothesis. In 1933–1937 was a director of the Physical Institute in Kharkov.

Leontovich, M. A. (7. 3. 1903–1989)
Physicist, theoretician. The founder of Soviet scientific school in radio physics, plasma physics and thermonuclear fusion.

Lifshitz, E. M. (21. 11. 1915–1985)
Physicist. Made contributions to a wide range of theoretical physics.There is a Lifshitz criterion in the theory of phase transitions, the Landau-Lifshitz equation in ferromagnetics, and others. Together with Lev Landau wrote the famous "Course of Theoretical Physics". Academician. State prize (1954).

Lunacharsky, A. V. (1875–1933)
Writer, scientist, member of the Academy of USSR (1930). One of the organizers of the Soviet system of education. Minister of Education (1917).

Lusternik, L. A. (31. 12. 1899–1986)
Mathematician and poet. A bright representative of Moscow mathematical school of Egorov and Luzin. In 1924 solved the Dirichlet problem by the method of finite dimensions. In 1929 together with Schnirelman completely solved the famous problem of Poincaré on the closed geodesic lines. Fundamental works on linear and nonlinear differential equations, on different aspects of topology. Published several books. Member of Academy of Science of the USSR. State prize (1946).

Markov, M. A. (13. 5. 1908)
Physicist. Academician. Main field the quantum field theory and high energy particles. Predicted several resonances which later were discovered. Found that the neutrino in beta-decay and those radiated by muons are not identical. Proposed (1958–1961) underground experiments for the study of high energy neutrinos and for seeking extragalactic neutrinos.

Mayakovsky, V. V. (1893–1930)
Poet. Painter. Actor. Started as a futurist. Then became the poet of the Revolution. Suicide.

Meyerhold, V. E. (1874–1940)
Producer. Actor. Theoretician of art of the theater. Created a new theater, with a new form of expression and decoration. Staged Mayakovsky's "Misteria Buff" (1921), "The Bedbug" (1929) and "The Bath House", Ostrovsky's "The Forest", and others.

Molotov, V. M. (1890–?)
Member of Communist Party from 1906. Member of Central Committee, President of Central Committee of Communist party (1926–1957). Fully responsible for all the distortions in internal and foreign policy.

Myasishchev, V.M. (1902–1978)
Aircraft designer, general engineer. Created a series of heavy bombers, such as M-2, Pe-2I, and others.

Nekrasov, A. I. (9. 12. 1883–12. 5. 1957)
Mathematician and physicist. Academician (1946). In 1930–1938 worked in "TsAGI" Central Aero-Hydrodynamic Institute, as a main theoretician in aircraft projects. Zhukovsky prize (1922), State prize (1952).

Noether, A. E. (23. 3. 1882–14. 4. 1935)
Mathematician. Created a new direction in algebra – the general, or abstract algebra, which strongly influenced mathematical thinking. In 1918 formulated one of the fundamental theorems in theoretical physics on the connection of the conservation laws with the symmetry of the system (the Noether theorem), which plays an essential role in classical mechanics, in quantum mechanics and in quantum field theory.

Petlyakov, V. M. (1891–1942)
Aircraft designer. One of the designers of the heavy bombers TB-1 and TB-3. Created the bombers Pe-8 and Pe-2. State prize (1941).

Pomeranchuck, I. Ya. (20. 5. 1913–14. 12. 1966)
Physicist, theoretician. The head of theoretical department in the Institute of Theoretical and Experimental Physics in Moscow (from 1946), the head of theoretical department in Kurchatov Institute and in the United Institute of Nuclear Research in Dubna. Academician from 1964. Created a new branch of theoretical physics – the physics of super-high energies. Achieved fundamental results in solid state physics. In 1950 predicted the effect in ^{3}He at low temperatures which carries his name.

Schnirelman, Lev (14. 1. 1905–24. 9. 1938)
Mathematician. Professor of Moscow University (1929). Together
with Lusternik solved the Dirichlet problem on closed geodesic lines.
Opened a new direction in mathematics - metric theory of the consequences of numbers. Solved the Goldbach problem. Left a rich heritage in many fields of mathematics. Suicide.

Seaborg, G. (19. 4. 1912)
Physicist and chemist. In 1942–1946 was head of plutonium part of
the Manhattan project in Chicago University. Led American research
into isotopes. Participated in discoveries of transuranium elements.
Nobel prize for chemistry (1951).

Sinelnikov, K. D. (29. 5. 1901–16. 10. 1966)
Physicist, member of the Ukrainian Academy of Sciences. In 1928–30
worked with Rutherford. Since 1944 director of Physical-Technical
Institute in Kharkov. Pioneer of accelerators and experimental high
energy physics.

Stechkin, B. S. (1891–1969)
Mathematician and engineer, member of Soviet Academy of Sciences.
Creator of the theory of thermal air craft engines. Pioneered and cre-
ated a series of jet engines. Hero of USSR (1961). State prize (1946),
Lenin prize (1957).

Sukhoi, P. O. (1895–1975)
Russian aircraft designer. Twice Hero of Soviet Union (1957, 1957).
Designer of jets Su-9, Su-15, and others. Created supersonic fighters
with arrow-shaped and delta wings. Aircraft T-431 and T-405 are
world record-holders for altitude and speed.

Suslin, M. Ya. (15. 11. 1894–1919)
Russian mathematieian, creator of the desorptive theory of sets. Dis-

covered the A-sets which appeared to be the Borel sets. Left rich mathematical heritage: Suslin criteria, Suslin theorem, Suslin's condition, Suslin's number, and others. All these results were described by Nikolai Luzin in his book "Lectures on Analytical Sets" Suslin himself published only one small note which became the basis for the famous work of F. Hausdorf "The Basic Theory of Sets".

Tamm, I. E. (8. 7. 1895–12. 4. 1971)
Physicist, theoretician. Pioneer of several branches in modern physics. Built the complete theory of the scattering of light in crystals. By the methods of quantum theory discovered the specific energy levels which opened a new field in physics – quantum theory of photoeffect in metals. This discovery played an essential role in the development of modern electronics. Created the famous theoretical school (based at the Lebedev Institute in Moscow). Hero of Soviet Union. Several State prizes. Nobel prize (1958) for the Vavilov-Cherenkov-Tamm effect.

Triolet (KAGAN), Elsa (1896–1970)
French writer. Born in Moscow. First novels and stories in Russian. Last husband Lui Aragon.

Tukhachevsky, M. N. (1893–1937)
Marshal of Soviet Union (1935). Published works on military science, organizations and building. Arrested and shot.

Urison, Pavel (3. 2. 1898–17. 8. 1925)
Mathematician. Pioneer of topology. Fundamental results in several fields of mathematics. There are Urison equations, metric theorem of Urison, lemma of Urison, the Urison space, and others. Created the theory of dimensions. Drowned while swimming off the coast of Brittany during a storm.

Vavilov, S. I. (24. 3. 1891–25. 1. 1951)
Physicist, academician. The President of Soviet Academy of Sciences. Founded Physical Institute in Moscow (Lebedev institute). Discovered together with his student Cherenkov effect which led Cherenkov, Franck and Tamm to Nobel prize. Fundamental experimental results and technical solutions in optics. Wrote several excellent books popularizing physics. Author of several books on the history of physics ("Newton", "Lomonosov"). The main editor of the Great Soviet Encyclopedia.

Valter, Anton (24. 12. 1905–13. 7. 1965) (in Chapter 12)
Physicist, member of Ukrainian Academy of Sciences. Fundamental results in experimental high energy physics, vacuum techniques, solid state physics. Built the first large accelerator in Europe. Together with Leipunsky, Sinelnikov and Latyshev produced the first fission by artificially accelerated particles.

Valter, P. A. (1888-1947) (in Chapter 14)
Mathematician and engineer. Fundamental works in hydrodynamics and in the application of aeronautics. Worked on Soviet aircraft projects in "TsAGI" with Tupolev.

Yumashev, A. B. (1902)
Test-pilot, air force general (1943). Hero of Soviet Union. Was in Gromov's crew in nonstop flight Moscow – USA (1937).

Zeldovich, Ya. B. (8. 3. 1914–1988)
Physicist, theoretician. Pioneer of the physics of reactors and nuclear energetics. Together with Sakharov and Khariton was the main theoretician in A- and H- bomb projects. Fundamental results in physics of neutrinos, high energy elementary particles, cosmological magnetic fields, and others. Three times Hero of Soviet Union. Four State prizes, Lenin prize.

Zheltukhin, N. A. (1915)
Mechanicist and engineer. Fundamental work in gas dynamics and thermal processes in different kinds of engines. Worked for many years with Korolyov on jet engines.